Becoming Top 1%

Becoming Top 1%

The code of engineering

By Sahin Boydas

PPS
POST PRODUCT STUDIOS

This book is a work of non-fiction based on the author's experiences and insights into the field of software development, its historical context, and its impact on our future. While the narratives and discussions are inspired by true events and personal experiences, care has been taken to respect privacy and confidentiality.

ISBN: 978-1-961954-05-2

Publisher: Post Product Studios

Printed in the United States of America

1603 Capitol Avenue, Suite 413A 2526
Cheyenne WY 82001

For more information, please visit www.postproductstudios.com

For inquiries regarding permissions, please contact:
publishing@postproductstudios.com

About the Author

Sahin Boydas, a Turkish entrepreneur who has carved a unique and successful path in Silicon Valley, stands as a shining example of what can be achieved through determination, innovation, and a deep understanding of the tech landscape. His journey from Turkey to the heart of the global tech industry is not just a story of entrepreneurial success; it's a narrative that resonates with the dreams and aspirations of every immigrant who has ever dared to envision a better future in a new land.

Before rising to prominence as a serial entrepreneur, Sahin began his career as a developer, immersing himself in the world of coding and software design. This foundational experience in the trenches of technology not only provided him with a robust technical skillset but also instilled in him a profound respect for the craft and creativity inherent in the field of development. This background has been instrumental in his success, allowing him to seamlessly integrate technical expertise with visionary business strategies.

Sahin's entrepreneurial journey is marked by the creation of four groundbreaking startups. From MovieLaLa, a trailblazing movie marketing platform, to RemoteTeam.com, a comprehensive solution for remote work complexities, his ventures have consistently pushed the boundaries of innovation. His success culminated in the acquisition of RemoteTeam.com by Gusto, a landmark achievement conducted entirely over Zoom, showcasing his adaptability and forward-thinking approach in the modern business world.

An alumnus of prestigious accelerators such as Stanford StartX and 500 Startups, Sahin's commitment to continuous learning and growth is evident. He has shared his expertise and insights as a sought-after speaker at over 400 conferences and podcasts and as a contributor to renowned publications like Forbes.

With a social media following of over 2 million, Sahin has established a significant online presence, engaging with a diverse global audience and sharing his journey and insights across various platforms. His influence extends across linguistic and cultural boundaries, engaging communities in both English and Turkish.

The decision to author a book stems from Sahin's desire to share his rich experiences in navigating the tech world, building successful startups, and embracing the future of work. His book is envisioned as a source of inspiration and guidance for aspiring entrepreneurs and business leaders, illuminating the path through the challenges and triumphs of the startup ecosystem. Through his narrative, Sahin aims to reach beyond the confines of technology and business, touching lives with his story of resilience, innovation, and transformative leadership.

Complementing his entrepreneurial ventures, Sahin has embarked on an ambitious project to educate 10,000 developers. This initiative reflects his belief in the transformative power of education and his commitment to nurturing the next generation of tech talent. By sharing his knowledge and experience, Sahin is empowering a new wave of developers with the skills and perspectives needed to excel in the dynamic tech industry.

Further extending his educational impact, Sahin has launched "The Becoming," a program designed for individuals seeking to enhance their way of thinking and perspective on life. This program encapsulates his experiences and learnings, offering a transformative journey for those looking to evolve personally and professionally. More about this life-altering program can be explored at The Becoming Institute.

Sahin Boydas's story is a testament to the possibilities that unfold when ambition meets opportunity. As a Turkish entrepreneur who has made a significant mark in Silicon Valley, he embodies the spirit of innovation and the relentless pursuit of excellence. His book and educational initiatives are extensions of his journey, offering wisdom, inspiration, and guidance to those who dare to dream and innovate, regardless of their origin.

Introduction

A respected occupation has continued throughout history, changing to fit the current era. Although the title may have changed every era, developers are the builders of history.

While we call these individuals "developers" today, there is an ongoing discussion about whether society has truly entered the era of genuine software development.

Although considering oneself a developer does not guarantee expertise in the field, a developer should always be open to learning and transitioning. This realization marks a significant milestone in a developer's historical journey and the growth of their consciousness to reach their destination. With this understanding, software developers can refine their skills and recognize their vital role in shaping the constantly changing world.

We can better understand the vital role of developers through the various repetition of events throughout history. Each repetition of these inevitable returns creates parallels with previous versions, leading to cycles. Societies often integrate previous ideas with new discoveries, thus creating, not a straight line, but a series of interconnected curves or loops. As long as those curves maintain their natural progression, they generate technological advancements.

In essence, however, this process was not about creating something entirely new, but rather the continuation and transformation of what came before. Today, we refer to this phenomenon as innovation, although it has already moved beyond that stage.

The process of innovation started with pioneers who observed their surroundings and shared their knowledge with others. They acted as an early version of Google, imparting vital information to their communities and cultivating new ideas in different fields.

Those historical phases have elevated humanity from being pioneers to becoming designers. However, the true potential of software development has yet to be fully unleashed.

The potential of software development lies in its ability not just to solve today's problems but to predict and address the challenges in the future. Imagine a world where software can enhance human capabilities and think empathetically, making decisions that benefit individuals and our society. We're on the cusp of developing programs to understand human emotions, simulate complex real-world scenarios in virtual environments, and even bridge the gap between our biological and digital existences.

When considering all this information, it becomes evident that developers play the role of architects in shaping the trajectory of history. At this juncture, there is a crucial point to emphasize: to make progress, it is necessary to have a thorough grasp of history and awareness. In order to comprehend history, consciousness is essential, as it represents the true understanding of a subject. To fully grasp this level of comprehension, we must categorize consciousness.

In reality, a developer designs and puts into action existing concepts. Therefore, they must recognize their role in this historical process as individuals who contribute to making history.

Building on this groundwork, we will examine the interaction between these two concepts, unveiling the developer's level of consciousness. In the process, we will uncover a vast library. Within this library, we will discover the meaning of consciousness, followed by examining our structure of consciousness. Since the starting point is rooted in history, history will become inseparable from our existence.

As time passed, the complex operational role of pioneers shifted into design with the changing habits of society and fast advancements in technology. The internet's communication network—one of the newest advancements in worldwide history—made all that possible. After this transformation, society embraced remote work.

When curiosity about events occurring on the other side of the world emerges, there is no longer a requirement to undertake a journey by ferry or train to explore. With the advancements in technology, and the new working culture that came with it, designers have already closed that distance. They represent the latest generation of desiring programmers, embodying the fundamental connection between design and history.

The practical designs of the wheel will change, becoming entirely new forms. We will see many examples in this book in that far-off future. While some might think that each experience will change based on future designers' lives, it is today's designers who will push society to that advanced level of software development.c

This book aims not only to instruct us on how to become a better developer, but to illuminate the principles for realizing our full potential as a developer.

Many strive for greater understanding, but it is necessary to unravel history to achieve such a state. Once we truly understand this process, we will unintentionally gain insight into ourselves.

As a result, after finishing this book, you may need to go back to the beginning, read again, reassess, and explore further. In this book, not only will we uncover the core of our profession, but we will also discover our identity. Just as personal growth is a continual process, engagement with this book is similarly iterative.

This recommendation is not a marketing tactic, but it promotes a relaxed approach to rediscovering oneself throughout their journey. While some readers may value annotating as a form of engaged reading, others might discover that this book unfolds its nuances over multiple readings, similar to a beloved film that reveals new insights each time you watch it.

By the end of this book, you will realize that the software development process is just starting. We will explore this well-established occupation further and carry forward its enduring impact throughout history's records.

The fresh perspective that developers have brought to the historical process has altered our levels of consciousness. In recent times, people have navigated between stages of individual consciousness, collective consciousness, and taught consciousness. Therefore, we must explore how these notions impact software development and technological evolution.

Get ready for a narrative where the past, present, and future converge, offering insights into the future of our digital world.

The beginning stage, the past, involves taught consciousness in the present time and putting forth desired values. Taught consciousness implies that our values and beliefs are not derived organically or through personal experiences, but rather taught or imposed upon us. Individuals, societies, governments, and nations express their preferred ways of living, imposing their desires.

Those desires spark worldwide discussions, leading to either acceptance or refusal. Occasionally, conflicts emerge due to such refusals. Ultimately, the desired results become evident. That is the condition taught consciousness instills to realize those desires.

Although we haven't discovered them yet, like the limitations inherent in matter, consciousness also has specific boundaries. Within those boundaries, individuals transform and reconfigure pioneers' outdated values into desired ones.

The collective consciousness of those pioneers' communities begins to display learned values. Having navigated various business landscapes, pioneers absorb those learned values and naturally incorporate them into their professional conduct. Consequently, the pioneer then imparts the art of conducting business to the merchant. That guidance arises from the accumulated and desired values.

The pioneers of the past and designers of today exhibited their individual consciousness, their personal motivations and ethics, within society's accepted values until the emergence of the internet. Before the digital age, innovators adhered to societal norms and values.

Before the internet, individuals' abilities to impart change, innovate, and influence were often constrained by physical boundaries, access to resources, and the prevailing informational and technological infrastructures. Personal consciousness—an individual's cognizance, moral, and ethical considerations—

often operated within those constraints. However, with the rise of the internet and the innovative design perspective of pioneers, the process changed. Transformative processes elevated personal consciousness to the desired level.

That suggests the trajectory started to change—shifting from localized, physically bound operations and influences toward a more globally interconnected and digitally mediated paradigm. Nevertheless, it is important to recognize that the three-part structure—past, present, and future—remains constant, with only their order shifting. Thus, we enter the second phase.

The second phase signifies the current era, while the third phase reveals the necessity of becoming genuine developers within historical records and the future.

The interaction between individual, collective, and taught consciousness, as well as accepted, learned, and desired consciousness, determines the course of past, present, and future, in totality in our history.

Although one might presume that history accelerates or decelerates, the truth differs from such assumptions. The concept of history "accelerating" or "decelerating" often hinges on the perception of how rapidly, or slowly, events, developments, and changes occur within a specific timeframe.

Within that time frame, emerging values from the interconnections between three phases: past, present, and future rejuvenate technological concepts.

Desires have become dictated by individuals. The significant influence of Google on human consciousness has nurtured thoughts such as, "This may indeed be feasible," "A concept like this ought to exist," and "I have the ability to strategize and execute according to my own desires."

Fundamentally, taught consciousness transformed into learned consciousness as individuals acted upon their own desires rather than those of others. By conducting their own research on Google, individuals began expressing their personal desires regarding societal possibilities.

At first glance, that may seem like an expression of freedom. However, if an individual's personal freedom aligns with their newfound autonomy and they remain a designer, they will not attain the status of a genuine developer.

Prevailing expectations lead to the labeling of an individual as a developer. However, as emphasized in this book, we have yet to attain the phase of genuine software development. To be a genuine developer, one must possess a natural individual consciousness and maintain it at their preferred level.

Collective consciousness has ceased to be a learned value and has transformed into an accepted one. This is because society employs the designs created. Embracing the designs, society favors the accepted instead of the learned.

During the acceptance phase, individuals may express their like or dislike of something, but that does not signify rejection. Individuals make choices based on their current values and the established order.

While an individual's choices might mean rejection, it does not necessarily imply that they have not learned from the experience. Through this process, individuals elevate the accepted value to the realm of taught consciousness.

Those who previously occupied the role of being taught shift into the role of learners. The act of being taught is a passive stage, focusing only on receiving knowledge shaped by external entities and conforming to established paradigms without critical engagement. In contrast, "learners" transform their engagement with knowledge. They become active, exploratory, and critical in their approach. Learners seek, evaluate, and co-create knowledge by interrogating information, navigating diverse sources, and synthesizing new insights.

If we wonder, "What comes next?" the answer is that collective consciousness will rise, not to the desired level, but to the learned level. That implies teachers will transform into learners, paving the way for developers to initiate a historical shift.

At that point, a disruption will take place in the process. All elements will undergo change. As previously discussed, the transition will progress:

From accepted individual consciousness to desired individual consciousness. This is the phase where individuals evolve from merely accepting the status quo and established values to actively shaping and aspiring toward a more refined, desired consciousness. Their personal evolution, in turn, plays a foundational role in shaping collective beliefs and the wider educational paradigm.

From learned collective consciousness to accepted collective consciousness. As individuals undergo their personal transformation, it creates a ripple effect on the collective. Societies, influenced by the aspirations and desires of their members, move from a phase of assimilating and internalizing knowledge (learned consciousness) to recognizing, endorsing, and accepting these new values and understandings as the new norm.

From desired taught consciousness to accepted taught consciousness. The educational or instructional paradigms, initially geared toward instilling specific desired values, undergo their own evolution. As societal and individual values shift, so do the teachings, aligning them with the new "accepted" norms. What was once a "desired" outcome of instruction becomes the universally "accepted" standard.

While I listed these transitions sequentially for clarity, they are inextricably linked, forming a continuous feedback loop. Each step impacts the subsequent one, ensuring that individual transformations influence collective beliefs, which in turn reshape educational paradigms. In essence, these aren't just transitions; they are a harmonious dance of evolution across individual, collective, and instructional spheres.

Navigating the process across three distinct levels of consciousness, we will recognize one's role in shaping history. The Latin roots of a news process elegantly express this consciousness.

We can visualize the Latin approach to the concept of "new" through its word for new, which is *nova*. Nova, translating to new, implies not just the dissemination of previously unknown information but the replacement or alteration of old information with fresh insights or perspectives. In this framework, news isn't simply an addition to what we know; it actively transforms the informational landscape, just as evolving values transform collective consciousness.

Indeed, consciousness mirrors the process of news dissemination. Individuals shape this process. They introduce new values to the news, subtracted from the realization phase, and made utilizable.

Therefore, the necessity to improve consciousness becomes apparent. This necessity will form our journey. Along this trajectory, we will learn to refine consciousness. Acquiring knowledge is essential, but success depends on embracing that knowledge for personal development.

The manifestation of our abilities doesn't come solely from acquiring them. Instead, it stems from accepting them and finding new ways to use them. This book won't provide you with entirely new knowledge; rather, its purpose is to help you rediscover the knowledge you already have.

You will either inhabit history, bear witness to history, or shape history. Right now, you find yourself amid history, and the situation will continue to be so. It is crucial that you deeply engage with history, as such intimate involvement will guide the advancement of consciousness.

A revelation awaits you: you persist in shaping history. Conflicts among nations no longer determine the defining moments of history. Instead, the transition of a developer from the role of a designer to that of a software developer does.

The alteration of the previously mentioned processes, notably the transition from the designer to the software developer role and the adjustments in consciousness phases, may appear to be a straightforward optimization process. However, we are not singularly engaged in mere optimization. Instead, we inhabit an authentic existence.

Chapter 1 : / ... / Historical Evolution and Modern Impact

A close examination of history is crucial to understanding the connection between software development and history, as it illuminates the relationship with remarkable clarity.

You probably know that structures like the Egyptian pyramids, Roman cisterns, and Cappadocian underground cities required a level of mathematical and geographical knowledge that surpassed what people commonly knew in their times.

However, there is one aspect we often overlook. When creating their plans, those civilizations had to include knowledge about the stars and space.

The builders demonstrated advanced architectural and mathematical expertise, and their profound understanding of astronomy is evident in the remarkable alignment of the Great Pyramid of Giza with precision to the cardinal points of the compass. Furthermore, the pyramid's sides closely correspond to the three stars of Orion's Belt, a configuration of great importance to ancient Egyptians.

Studying the tablets, writings, and structures from those times shows their notable progress was based on astronomy. And a society deeply engaged in the study of astronomy would naturally have an insight into mathematical and geographic algorithms.

Without that understanding of algorithms, those societies would not have been able to calculate spatial values accurately using only the naked eye. Nowadays, this task requires advanced telescopes.

The interest in space and the effective application of their geometric systems to various geographies would naturally have needed a software-based process. Historical records not only reveal the creation of civilizations by humans, but also their involvement in both the rise and fall of these societies.

One of the most important additions to the advancement of software development was made by the father of robotics, Al-Jazari. A versatile inventor from twelfth-century Mesopotamia, his development of complex water clocks and intricate automata that used principles of early programming is a testament to his prowess. His achievements as a mechanical engineer, artisan, and mathematician provide valuable knowledge to this day.

It is worth mentioning that software is inherently connected to every phase of historical advancement. And there is a possibility that the catastrophe of war could destroy our modern technological progress.

In such a scenario, around three to four hundred years from now, people might perceive the software methods of our time similarly to how we view hieroglyphs.

Therefore, developers should develop a new perspective on historical works within the framework of this algorithm.

This indicates we should go beyond traditional mathematical and logical patterns, reevaluating our studies on Egyptian pyramids by using the viewpoint of spatial science algorithms.

An obstacle for developers in carrying out this assessment is their insufficient understanding of the methods to approach history. If you think you are the final observer in examining history, you are mistaken. This is because every person who makes a discovery through scientific methods knows they are at the beginning, not the end.

In short, this is the start of a new adventure, and announcing it is just the beginning, not the end.

New discoveries can emerge by adopting a fresh algorithmic perspective, embracing its components, using innovative software methods, and taking a pioneering approach. We will re-evaluate these aspects within the range of possibilities. The mistaken belief that we have reached the highest point of our understanding hinders our view of history through this perspective.

The technological advancements of the last ten years, including displays and mobile phones, signify the start of a changing period in our lives, one that triggers ongoing development. When we examine the development of these values within the context of civilization theories, the continuous process of progress becomes apparent.

History does not move in a straight line; instead, it cycles in balanced circles, rotating around a point of change. Although confined within a cube, history constantly repeats itself. When repetitive processes influence and complete each other, new technological advancements emerge. To advance in technology, it is essential to compare the realms of health, medicine, sociology, economics, and politics within historical civilizations and examine their reflections in the works of those societies.

A formidable challenge confronting modern developers is the ongoing demand for optimization. However, the reality is that optimization is an unattainable objective in a continually evolving technological landscape. Therefore, our endeavors in its pursuit are mere self-delusions.

At the heart of the issue lies the capacity to face reality—a reality that recognizes history as an inclusive sequence of the past, present, and future. The most tangible entity in this sequence is history and our position within it. Upon recognizing our role in history as a continuous sculptor and key participant in the process, we will comprehend the perspective of this consciousness.

Consider immersing yourself in historical literature to deepen your understanding of your consciousness. As you delve into those texts, envision the narrator's perspective and observe the past through your present lens. Now, reflect this knowledge on the news. Changes in the instruments and weapons used in warfare do not modify historical realities. You may discern similar patterns in the present, illustrating that the motives and mechanics of wars throughout history have largely remained unchanged.

It is essential to acknowledge that we exist both as a participant in history and as its creator. As a result, the designer's role will elevate to that of consciousness. Once we establish consciousness, we will witness the emergence of a genuine culture and a professional approach to development.

Underestimating this is not advisable. Once we fully grasp the authentic essence of software development, it becomes our responsibility to acknowledge the era of the designer in which we currently find ourselves.

When we define "consciousness," we're referring to how developers must reflect on their actions and motivations throughout their lives. It's a process they must go through to understand themselves and their existence better. If one cannot understand their actions or the underlying reasons, mere "knowledge" will indeed prove insufficient, which is already evident in the present era. The key does not lie in a developer's possession of knowledge, but in their recognition and acceptance of it.

For example, you may currently be engaged in developing a chat program similar to WhatsApp. However, imagine the implications when you realize that the equivalent of WhatsApp was carrier pigeons and other such creatures in the past. Would your perspective change if you recognized that horses and wolves were once instrumental in communication processes? Upon comprehending the harmony with natural processes, a topic elaborated throughout this book, you will realize that you are reconstituting that reality in the present.

We continue to communicate using a variety of tools. However, the diversity of these communication tools does not imply that we have progressed from the design phase to the software development phase.

One might argue, "Look at the computers; they are constantly engaged in coding." However, the development process far exceeds the mere confines of design. While it may be true that we currently lack the technology to surpass the design phase, such technology is on the horizon. Therefore, we must prepare ourselves for this impending advancement.

Chapter 1: /... /... / The Art of Words in Deepening Consciousness

Let us revisit the initial discussion about the journey toward comprehensive understanding in our field. We should orient our pursuit toward understanding rather than a mere accumulation of knowledge. However, we must be aware of a subtle distinction: the boundary between acquiring and acknowledging knowledge. These concepts share a close relationship and often overlap. The relationship and differences between knowing and understanding can often be elusive for many.

When words possess similar meanings, ambiguity is bound to arise. In the present context of software development, one of the most prevalent sources of ambiguity stems from the fact that codes are often overly similar, narrowly specialized, and excessively constrained. Despite writing millions of lines of code, you continue to employ and reshape identical terms. Paying attention to the similarities of those terms, which you clearly recognize, can inadvertently impact your process of acknowledging their nuances.

Because of that influence, you may use different words that mean the same thing. As a result, you exist within a limited perspective where those synonyms lose their meanings. Although you may be aware of this phenomenon, you might not realize that your actions have an impact across multiple dimensions, potentially as many as thirteen at the moment.

By going back to the source or origin, that situation limits your ability to create a distinct language for software development, which, in turn, obstructs your path toward achieving truly innovative results.

Consider this: there are no absolute synonyms in our world. The belief in their existence arises from our limited daily routines, compounded by the flood of words, concise expressions, and emojis offered by what we perceive as advanced technology. Historically, each word carried a distinct implication. Therefore, immerse yourself in the realm of words. A good starting point could be Latin. As you delve into the etymology of words and recognize the absence of absolute similarities in meanings, you will discover that while lines of code may have identical values, they manifest as distinct combinations with different functionalities in various contexts. Subsequently, you can embark on a new journey in the realm of development.

The creation of combinations require a level of depth, which is nurtured through our playful exploration with words—our fundamental building blocks—as we shape our consciousness. As a result, we can say that our initial practice as developers is to engage in linguistic play.

> Exercise_

At least once during your day, choose a word and embark on a linguistic exploration. List all its synonyms and delve into their connections with Latin, French, Spanish, and other languages, exploring their etymological roots. Engage in this practice for at least a few weeks or until you notice patterns and relationships among the words. By tracing those roots and examining their usage, you will discover that authors or speakers of the past attributed varying interpretations.

Perceiving different words with similar meanings is comparable to observing a tree from a distance. The inability to partake in its fruit enhances the value of the unattainable seed. Indeed, the seed of a fruit contains an entire tree within its compact form.

History and nature provide remarkable insights, allowing developers to grasp the algorithm of values truly. When you gaze upon a tree adorned with thousands of leaves, you perceive only the abstract idea of leaves. Nonetheless, upon meticulous analysis and categorization, you will realize that no two leaves are identical. Their functions may even differ from one another.

Just as our organs and their corresponding systems generate new functions and unexpected symptoms through enzymatic interactions when operating logarithmically… And as our stomach, primarily responsible for digestion, can abruptly shift its function to eliminate toxic substances by expelling them through the act of vomiting…

Ultimately, engagement with words requires verbal discourse rather than solely written communication, and associations emerge through wordplay, which stimulates brain waves.

In today's world, developers may think they are using their mental capacities to the fullest extent, given the surplus of tools, platforms, and resources providing instant access to extensive information and problem-solving techniques. However, engaging in dialogue rather than relying solely on written communication is essential to achieve optimal cognitive efficiency. While we have resorted to writing when interacting with distant acquaintances and have even elevated it to a cultural activity, in the long run, it can potentially disrupt our neural wave patterns.

Do not forsake the art of conversation.

> Exercise_

Create moments to share ideas and conversations with others. When you engage in discourse, remember there are no absolute synonyms, so enrich your lexicon and use new terminology. After using new terminology and improving your sense of understanding and communication, others may question, "Why use such an old term now?" or "What does that mean?" Do not hesitate to provide an explanation.

The act of explanation is vital, not only for influencing others, but also for your personal development. Merely repeating information to oneself does not lead to fruitful results. Your neural wave patterns evolve through interactions with others, thus enhancing your consciousness.

Currently, many software programmers have limited brainwave activity, and their consciousness fails to progress. Consequently, they distance themselves from evolving into the developers of the future.

Within the realm of verbal communication, two elements hold prominence: the message and the word. The message carries energy and retains a personal character, while the word embodies a shared energy exchange.

Every word you utter sends out sound waves created by the vibrations of your vocal cords, which have the potential to impact and modify the environment and interactions around you. Every word you articulate alters the circumstances through the frequency your vocal cords generate. That doesn't mean speaking a word can literally change physical circumstances. However, words can indeed influence the social, emotional, and psychological dimensions of a situation.

As mentioned before, the absence of synonyms present another challenge. However, it is important to highlight that two words having different meanings at different points in time are a linguistic reality arising from variations in sound frequencies.

In the present day, we can evaluate our personal development by examining the sound frequencies of similar words spoken at different times. In the future, when humans find themselves interviewed by artificial intelligence (AI) for job interviews, computers will not only differentiate between truth and false, but also perceive the genuine belief embedded in spoken words.

Therefore, when you state, "I want to work in this company," the emphasis on a specific word becomes irrelevant. Even in cases of attempted deception, the frequency produced by your vocal cords would not allow any aspect to go unnoticed.

Another undeniable truth is that trees are highly aware of the approaching winter. If every tree on Earth could understand us, and we informed them about their upcoming loss of leaves in autumn, they might indeed express objections. However, the reality remains unchanged.

Humans, like trees, seem disconnected from this level of consciousness, operating under the belief that they can maintain their awareness solely through personal communication in their daily lives. If we view ourselves as developers who make significant declarations within the realm of communication but remain detached from personal interactions with others, we will inevitably limit our potential for progress.

Energy serves as the fundamental basis for communication, but it needs to be transmitted to become a part of us. If you don't actualize the conveyance of energy, there's a risk that it might interrupt your process of interpretation.

As a result, one cannot achieve success in the field of coding without going through the process of interpretation. Attaining this success is crucial, not only for the benefit of society, but also for your personal gratification. To realize that satisfaction, it is imperative to engage with words continuously.

As a result of these linguistic exercises for meaning and consciousness, a new process will unfold, bringing with it an inherent risk.

In this book, we will analyze and break down concepts. Subsequently, we will clarify the influence of those elements when examined together with relevant historical and modern contexts, outlining the reasons behind their effects and detailing the strategies we should employ.

To illustrate this idea, let's reflect on your childhood.

Let's assume you had a persistent longing for a bicycle. You expressed your desire to ride one, recognizing its significance in allowing you to participate in games with your peers and experience a sense of inclusion in play.

The aspect of consciousness involved formulating the necessary arguments to acquire a bicycle, persuading your parents, saving money, selling some of your belongings to gather funds, and similar endeavors. As your mind became captivated with the concept of a bicycle, meaning and consciousness intertwined, giving rise to a new goal—a motorbike. That occurs because our brain experiences our imagination even before we embark on the next step. However, because of that, a complication arises.

This is one of the reasons developers make mistakes. Unable to fully integrate the meaning with their consciousness and distinguish between them, developers try to inflate balloons that have already deflated. This metaphor symbolizes efforts spent on outdated or irrelevant objectives, akin to attempting to revive a project or idea that no longer holds relevance or effectiveness in the current context. Much like a balloon that has lost its elasticity and cannot be restored to its original state, certain ideas or methods in software development may become obsolete or unproductive over time. The challenge for developers lies in recognizing when their endeavors are akin to inflating such a balloon – a task that, despite their skills and dedication, is no longer feasible or fruitful in the ever-evolving landscape of technology.

That parallels the challenges we face during our years as students. When presented with a question, we generally possess the solution. The answers are within us, and we subsequently connect those answers to the nature of the question asked.

When faced with an equation involving three variables, your consciousness reveals the methods for its resolution. As you understand the question, you merge meaning and consciousness in the pursuit of a solution.

Afterward, you might realize that your solution is incorrect, leading you to search for the mistake and try to rectify it. That highlights the most profound distinction between designers and other individuals. If you attempt to correct an error from its point of origin, your likelihood of making errors in regular procedures will increase. If you endeavor to address the question from its beginning, your mind suggests innovative approaches.

Embarking on a search for a solution from the source of the error itself induces stress. Despite the eventual possibility of finding the solution, this stress will consistently overwhelm you. Do not obsess over the temporal aspect during this process. While we will delve into this topic later, first, absorb this crucial point: analyze your mistake not from the error's junction, but from its very beginning.

You might wonder, "Do I need to return to the starting line? I have already programmed that, and it works flawlessly." In this situation, your consciousness strives to respond quickly, and your brain expands the pool of viable alternatives, then you use these options to cultivate a renewed consciousness. If you return to the beginning and start afresh, your brain will offer a heightened level of consciousness compared to your initial attempt. As a result, you will realize more errors and deficiencies.

Those errors and deficiencies, however, may not necessarily relate to the initial line you wrote. Similarly, upon examining the mistakes in your interpersonal relationships, you will notice a coherent pattern. As your mind revisits correct actions, it uncovers your mistakes. That poses one of the most significant challenges in the field of coding.

Indeed, Windows, developed and distributed by Microsoft, is an operating system software provided to us as individual and professional users. When users encounter errors, the software has a built-in mechanism that generates error reports, processes them, and attempts to resolve them. However, if the focus were on the processes leading up to the error rather than solely on the error itself, Windows would be a fundamentally different product today. It would be more versatile, advanced, and user-friendly and it would drive the process forward. This perspective is what drives the process of AI forward.

Let's consider another example. If our stomach rejects a substance, our body removes it. Instead of trying to control the issue once it arises, the body seeks to manage it from the very beginning. The level of control is so sophisticated that, if you come across the substance again, your body recognizes it by its scent, triggering a feeling of nausea.

It's not just your stomach that recognizes the substance. Your brain, eyes, ears, and even your thumbs participate in the process. Therefore, consciousness is a mechanism that can operate in our body with minimal errors and maximum efficiency.

Just like in the previous example, in the realm of coding, embracing a similar anticipatory and comprehensive error management strategy as our body's mechanism could significantly enhance software systems' robustness and user experience. It's also worth noting that fixating solely on correcting mistakes can often lead to a demotivated state and, paradoxically, further errors. By starting fresh and addressing our errors from the beginning, we not only boost our motivation, but we empower our brains to reveal new possibilities.

Consequently, this approach facilitates faster progress and enhanced efficiency in our developmental endeavors. Addressing errors from their origin might seem like a daunting task at first. However, the core issue resides in identifying the appropriate steps to begin with. In order to perform this evaluation, it is essential to revisit your coding lines multiple times and comprehend the steps taken thus far. That process eludes temporal measurement due to the complex nature of the coding process.

In our world, we conventionally define time within a 24-hour day. However, on Uranus, a day lasts only ten hours and fourteen minutes. This contrast highlights that the key issue is not about making comparisons based on time, but rather recognizing the existence and nuances of different environments. In software development, this means understanding that solutions must be tailored to specific contexts, not just based on a universal standard. By acknowledging these inherent differences, developers can create more effective and appropriate solutions, aligning their strategies with the unique requirements of each scenario. Thus, it's crucial to reassess actions and their validity in varied contexts to ensure relevance and efficacy.

Perhaps the solution is to examine small details to detect errors closely. However, we can analyze many of these details through simple observation and reflection, eliminating the need for microscopic examination.

In the pursuit of knowledge, modern exploration may require the use of high-definition electron microscopes to observe the movements of microscopic entities or atoms. However, investigation is not limited to the examination of atomic structures through electron microscopes. By observing the cosmos, one can perceive phenomena that mirror atomic behaviors.

The celestial objects within our solar system serve as an example of the dynamic interplay between atomic activities and orbital paths. Ancient societies analyzed atomic structures without the use of microscopes. They applied that fundamental approach and progressed to more advanced stages. Aside from that, they were fundamentally similar to us.

The ancient civilizations studied their environment and hypothesized that similar principles could apply at a microscopic level. They hypothesized that if they could refine and manipulate metals at a macroscopic level, some underlying atomic or molecular structures and interactions must be responsible for those properties. They recognized a connection between their cognitive processes and nature. Therefore, a simple observation of nature, without excessive analysis, was sufficient for them. In modern times, we often regard observing nature as a leisure activity or hobby. However, a thorough examination of the natural world can greatly assist developers in enhancing their comprehension and identifying mistakes.

Identifying significant differences does not always require the use of a microscope. However, that does not imply that we should completely disregard the tools or methods. There are numerous examples within the universe. Every ancient method might regain significance in shaping the future era. In essence, we can see every past method as new.

> Exercise_

Gradually, we'll delve into processes that combine history, consciousness, and meaning to aid the developer.

Before embarking on this journey:

- Identify your most recent error in life or at work and trace it back to its beginning.

- Start practicing from the point where you executed the task correctly, gradually approaching the error.

- On the same day, take the time to observe the stars, not through NASA imagery or videos, but with your own eyes. The metaphorical communication among the stars will provide a new level of consciousness and meaning, enabling you to rectify your errors. Their consistent patterns and rhythms will prompt you to ponder the nature of your coding practices, encouraging a similar consistency and logic.

Observing the cosmos is indeed one of the most influential and dynamic methods of harmonizing consciousness and meaning within the human mind.

For this reason, throughout history, individuals have observed the stars. However, they engaged in this activity for more than just predicting the future or confirming the existence of beings from other planets.

Spending an evening stargazing from your rooftop or, if you live in the city, taking a short trip in nature, will unintentionally create a link between consciousness and meaning.

Chapter 1: /... /... /... /... / Consciousness and Meaning: From Chaos to Harmony

As a developer, an essential part of your role includes attributing meaning to the ideas that arise in your mind. That requires gathering input from various sources, combining the gathered information, and developing a strong comprehension.

For instance, imagine you're conceptualizing a new app for facilitating remote work. Your idea might center around "seamless communication." Attributing the meaning of that idea involves dissecting the general idea into tangible features, like instant messaging, video conferencing, and file sharing. When imbued with specific attributes and intentions, each idea transforms into a roadmap that guides the coding and development process, ensuring the end product effectively embodies the initial concept in a functional and user-friendly manner.

Essentially, you are the one who assigns meaning to a project that various sources of information have shaped. Throughout the process, consciousness and meaning continuously interact with each other. They work together to reach a resolution by identifying detailed patterns.

That inherent interaction can inevitably lead to confusion because, as the developer assimilates multiple pieces of information, the convergence of diverse ideas, varied data points, and alternative solutions can blur the clarity of the original concept. As the developer, it is your responsibility to restore order amidst the chaos. Once you recognize your purpose, you will also comprehend the disorder that disrupts balance. The key lies in the cultivation of understanding and awareness.

As you explore those concepts within your realm and on a micro level, you can effectively respond to demands from broader contexts (macro meetups), turning them into valuable outcomes within a reasonable timeframe.

> Exercise_

Consider a proposal to update a main road by combining three, four, or more additional lanes, installing streetlights, and implementing changes to improve traffic flow.

The city's comity may identify the scattered intersections as the cause of traffic congestion. That insight arising from their awareness becomes a significant outcome that you, as a developer, must address. However, there is a secondary issue to consider: there may be ongoing maintenance or demonstrations taking place on that road. The client may suggest relocating or rescheduling those activities. That introduces a secondary meaning arising from another consciousness, which is the additional issues to address before taking the first issue. As a result, you find yourself involved in a project that involves multiple requests stemming from different perspectives. Therefore, avoid the urge to visualize the process immediately, which requires reinterpreting and contextualizing the requirements.

Such lucidity enables rapid and practical execution within the process, particularly within the domain of coding languages. Therefore, it is crucial to gain a comprehension of the confusion that arises during the accessibility phase. Confusion often stems from the incompatibility between our consciousness and meaning.

Imagine a student who takes a test in the morning. She has prepared herself well in advance. However, due to a stomachache, she becomes distracted during the exam and starts looking at her surroundings. Upon observing her behavior, the teacher assumes the student has not prepared. However, after considering the student's previous excellent academic performance, the teacher realizes there must be a more complex situation at hand. In this case, the teacher's consciousness prevents them from suspecting the student of cheating and reorganizes their approach.

Every obstacle holds significance as the starting point for a new arrangement. Indeed, obstacles often preserve our existence, but for a developer, they see each obstruction as the origin of a more complex design. As a result, additional applications will form increasingly complex setups within the system, prompting the emergence of new algorithms at the confluence of consciousness and meaning.

Since we are examining this from our personal viewpoint, it is appropriate to contemplate an instance with future technologies. Determining where to share your information among those diverse applications—such as decentralized finance platforms, secure communication apps, AI-driven health monitoring systems, wearable tech, and augmented reality interfaces—is crucial for your existence in the environment of future technologies.

In the near future, innovative applications will facilitate the simultaneous usage of similar applications. For example, you might develop software that allows users to edit photos with AI to morph into fantasy creatures, while other developers might work on the software to change backgrounds, provide live photo effects, or generate 3D object renderings. End users will then strive to reduce confusion and promote the integration of those software applications into a unified platform. Just as those innovative setups enable the use of multiple applications within a unified platform, unprecedented new methods to reduce confusion will continue to emerge.

Imagine this scenario: there is a car, bus, airplane, or spacecraft that can transport you from Los Angeles to New York. Each mode of transportation may intersect with another, potentially leading to confusion or delays. What if a software application could assist you in selecting the most suitable mode of transportation, ensuring optimal efficiency for a stress-free journey to New York? In such a scenario, you would undoubtedly be the most satisfied traveler.

When a developer creates code arising from the intricacies of a complex order, they are compelled to confront the natural realities. While those technologies simplify our lives, they also create complex setups that challenge the harmony of nature. Fields that address the harmony and resolve confusion will gradually become more important for software developers. Nature is devoid of confusion; it embodies transformation.

Interchanging elements assist and influence each other. If they impose constraints on one another, those limitations reveal new natural necessities while also supporting human existence.

We are at the edge of a new technological era where we will shift from a state of confusion to one of transformation. It is essential to face the complexities of our lives and transform them before embracing this upcoming era. Indeed, you may experience a challenging day, have an argument with your partner, receive negative comments from your family, feel disconnected from friends, and find yourself in a state of confusion. Transforming that negativity into something positive carries greater significance than anything else.

If you reconcile with your companion, you can experience a sense of renewal. While reconciling with your family can lead to personal transformation, it's crucial to approach their opinions with a balanced perspective. Before committing to any changes, take a step back and assess their views objectively. Ask yourself if these opinions are constructive and beneficial to your growth. It's important to recognize that while family perspectives can be valuable, they may not always align with your best interests or personal truth. By observing their opinions unemotionally and critically, you ensure that any changes you make are in harmony with your authentic self and contribute positively to your personal development.

For example, when a fifty-year-old man mentions it is his birthday, our response should reflect this transformative perspective: 'Yes, it is your birthday, but it is not just a year that has passed within the complexities of life. You have another year ahead, during which you can experience the richness of a thousand years.' This perspective underscores that the upcoming year holds the potential for many new experiences, learning opportunities, and personal growth. However, use this response judiciously, particularly with those you know well and who would appreciate a deeper reflection. In more casual encounters, a simpler acknowledgment may be more appropriate. If you truly believe in your words, you can inspire the aging man to make significant contributions to the beauty of natural balance.

In nature, none of its living beings operate with the awareness that they will eventually face mortality. That's why aligning ourselves with the natural order is essential, as it enables us to shift from confusion to transformation. The reason is that nature's transformation process doesn't involve any loss or deficiency. In case of loss, complementation compensates for it. Likewise, if there is a deficiency, it undergoes a transformation process.

For instance, if a tree in the forest dies due to old age or disease, it represents a loss in the ecosystem. However, in nature's transformation process, it compensates for the loss through complementation. The dead tree becomes a habitat for insects, fungi, and microorganisms. Those organisms break down the decaying wood, returning nutrients to the soil. The nutrients then nourish nearby plants, helping them grow stronger. So, even though the tree was lost, it contributes to the overall health and balance of the ecosystem through complementation. The point is that we can perceive the consistency that the world requires as a combination of completion and transformation, which, when working together, is what we know as harmony.

Is harmony not the expectation placed upon you within the realm of software development? Therefore, acknowledge that you, too, are involved in a procedure of completion and transformation. Despite the many challenges and experiences you have faced on your journey to becoming a developer, once you have achieved this status, you will continue to pursue that profession for the rest of your life.

It is important to realize that those experiences serve as complementing stages in a transformation process. Always remember that an event from your past lays the foundation for a future occurrence, ultimately shaping your transformation. As your memories and experiences come together, they will reveal a harmonious journey.

Every developer adept at harmonizing their memories will resonate with nature's conditions. By doing so, they will embark on an internal journey toward harmony. Thus, avoid making the same mistake as those who strictly pursue organization. Do not impose rules for yourself. Each rule you set becomes an obstacle to overcome. To achieve organization, you must comprehend the inherent rules within nature. To establish order, it is essential to understand the rules as they occur naturally.

Remember that the lion hunts its prey as instinct, not because it is aware of nature's rule. Its seemingly spontaneous yet inherent action underscores nature's implicit rules; creatures naturally adhere to the pattern or rhythm that facilitates survival and sustenance. Therefore, the principle of completion and transformation in nature is essential for life.

If you refrain from imposing rules for yourself and instead allow the outcomes of your harmony to naturalize, you can effortlessly adapt your own guidelines to your existence. To naturalize the rules set by your superiors, ask questions that delve deeper into the rationale and purpose behind them. Instead of asking, 'Why is it like this?' which might come across as confrontational or dismissive, try asking, 'Can you help me understand the goals we aim to achieve with this rule?' This approach not only demonstrates a willingness to understand but also encourages a more constructive and insightful dialogue. Such questions can lead to a better grasp of the transformative aspect of the rule, allowing you to align it more closely with your personal and professional growth.

When a rule enables transformation and simultaneously brings completion, harmony is accomplished. Remember, harmony is essential in managing a team. That principle serves as a cornerstone for leaders who desire to integrate the understanding into their personal and professional lives actively. Regardless of your rules, always examine what they metamorphose and complete.

If you make quick decisions without a thorough evaluation and set new rules based on those decisions, do not anticipate immediate compliance from your team. In a situation where individuals neglect to adhere to the rules, harmony is the first element to be affected, resulting in a gradual erosion of the principles of transformation and completion. As a result, losses and deficiencies emerge. At that stage, any attempt to impose something unnatural on the situation will only limit the completion and transformation of your team. This creates a new complex order for the team leader who transitions from leading a single process to cultivating this consciousness. Naturally, significant confusion will arise within your organization as you navigate through those stages.

Remember that rules are an inherent part of nature.

It is important to remember that nature does not easily accept the complexity we impose on it. In light of this reality, it is possible that developers will split into two completely different groups in the future. A significant conflict will emerge between those who support the ethics of nature and those who oppose them. The ethical values currently under discussion in the world are still developing within current level of consciousness.

However, as we endeavor to decipher this intricate arrangement using the principles of restorative mathematics, we come across unexpected results. This stems from the fact that no one has ever articulated such complexity in this way before. Developers shaping future technologies will engage in discussions regarding nature itself to address this complexity.

Applications that disrupt the natural order or go against nature will become more prevalent. Considering these advancements, we can divide the realm of software into two distinct categories.

In modern terms, the contrasting roles of white hat hackers and black hat hackers exemplify this division. Black hat hackers engage in illegal activities such as stealing funds, acquiring credit card information, blackmailing individuals with compromising content, or conducting other unauthorized actions for personal financial gain. White-hat hackers, on the other hand, are individuals who counteract the activities of black-hat hackers. They actively work to protect systems, identify vulnerabilities, and ensure digital infrastructure security.

We will always encounter the principle of duality. However, within the ethics of nature, what truly matters is not the simple division of good or bad, but rather whether one's actions are helpful or harmful to nature. At this point, supporters of this principle will go through, developing a deep respect for natural processes, such as adapting to nature. Developers who accept the ethics of nature will grapple with meeting the requirements, which include ensuring sustainability, reducing environmental impact, and aligning technological advancements with principles that support and enhance natural processes and ecosystems. That effort will confront the developers who seek to eliminate it, signifying that the historical progress of the software world will cycle through the "not good against bad, but bad against good" pattern, ultimately bringing us back to the beginning.

A phrase often attributed to Einstein, though not definitively, provides a sad reflection on this cycle: "I know not with what weapons World War III will be fought, but World War IV will be fought with sticks and stones."

To live a morally upright life, we must first understand the ethics of nature, as these ethical mechanisms are integral parts of our everyday existence. We strive to enhance ourselves and our consciousness through methods not aligned with nature, mistakenly believing that our consciousness is separate from this inherent aspect.

No matter what we do or where we are, we live on Earth, which consistently maintains its twenty-four-hour cycle. As time seems to pass faster with each passing year, it may appear to outpace the rate of technological progress. This feeling might stem from living a life that disregards the ethical principles rooted in nature. At this point, developers must embrace the ethics of nature to achieve genuine success. This dedication could very well be the primary motive for them to become future developers deeply committed to upholding the ethics of nature.

Embracing a lifestyle aligned with nature will cultivate developers dedicated to upholding nature's ethics. Living in harmony with natural principles fosters deep respect and understanding of its intrinsic values. This manifests in the technological creations and professional ethics of these developers.

So, how can we differentiate between a life lived in harmony with nature and one that deviates from it? The distinction becomes evident in the approach taken toward coexisting with nature rather than solely exploiting it for personal gain or survival. Suppose you align yourself with developers seeking to dominate and exploit nature for their ambitions, creating technology that heavily depletes natural resources and ignores sustainability. In that case, you inadvertently contribute to the pollution of the world. Nonetheless, it is important to acknowledge that nature possesses an inherent ability to self-cleanse and regenerate.

When contemplating the connections between oceans, forests, and other natural features, the capacity of nature to cleanse itself effectively becomes evident as long as it progresses without interference. In a similar way, individuals embody the same principles. If a person encounters a challenge and allows themselves to surrender to the natural rhythms of life, understanding that the issue is part of their journey of transformation and completion, their existence tends toward natural healing.

It is not feasible, though, to resolve difficulties solely through excessive contemplation. Continuously dwelling on problems hinders finding solutions. The crucial point is not to comprehend why an event occurs, but rather to understand its consequences. As a result, the most crucial active role in the life of a developer who embraces the ethics of nature is not to analyze the root causes of problems, but to swiftly implement solutions that mitigate the impact and prevent further issues.

Upon recognizing the effects of an event, an individual fulfills the requirements of being human. Subsequently, they recognize the importance of being part of nature. Armed with that new awareness, black hat hackers expect opposition to nature will become less clear-cut, instead, changing into a more ambiguous gray area. In a world filled with sharp divisions and absolutes of black and white, an unavoidable spectrum of gray areas emerges in between. The spectrum represents developers who swing between black and white, driven by their personal interests.

We cannot achieve embracing nature solely through further evaluation or analysis. Therefore, it becomes essential to instigate change in our everyday lives. We should recognize our role within nature and thoughtfully assess the surrounding processes. Considering those ethical values, we must understand that humans do not participate in an ongoing fight for survival; instead, we simply live in the natural flow of life. If individuals view their existence as an ongoing struggle for survival rather than a simple natural state of being, it is likely to result in excessive stress.

Avoid overexerting yourself in the pursuit of becoming a developer; instead, immerse yourself in your work in software development. If individuals try to force themselves into a specific role, they will not receive support from nature, as nothing in nature consciously strives to "become" something. Those who give their best always receive their reward in nature, as they find a kind of serenity and equilibrium that comes from living in harmony with the natural world. The idea of individuals achieving superiority, in terms of social status, wealth, or intellectual prowess, through concentrated effort remains unproven and instead represents a deeply ingrained artificial reality.

Avoid being captivated by the artificial world. The tangible reality of your physical existence will disrupt the things you seek to comprehend through artificial means, leading you in unexpected directions. Those complex paths will muddle your perception of the world. From coding to designing, each stage will gradually become more burdensome and inefficient. Your responsibility is to focus on what you believe is right. Your efforts will naturally integrate you into the processes of completion and transformation. Staying consistent in this journey will lead you to a point of holistic transformation.

Climbers reach the highest point of mountains; however, it is not only their physical effort that guides them to the peaks. It is their unwavering commitment to taking the proper steps from the beginning that drive them toward the top.

Walk. Simply walk.

No one can reach the top by running.

Chapter 2: /... / Transformation Through Disruption

As we strive to uphold the ethics of nature, let us first evaluate the current situation and examine how nature achieves balance. By undertaking this task, we can incorporate these values into our lives, leading to successful outcomes through their implementation. For instance, let's consider the process of nitrogen transformation.

Nitrogen plays a vital role in our world; it is the most abundant element in our atmosphere. The solvating effects of lightning facilitate the transfer of nitrogen from the atmosphere to the soil. This process occurs through thermal effects and disruptive factors. Storms serve as conduits for nitrogen transfer, while decomposing remains add nitrogen to the soil. Lightning not only dissipates excess electrical charge from the soil, it also eliminates bacteria in areas rich in nitrogen resulting from decomposition.

As storms carry these elements, they create favorable conditions for the growth of seeds made up of nitrogen bonds. Afterward, the following spring brings a variety of unexpected flowers to our garden. Therefore, we observe a flower we have never seen before, blossoming due to the combined effect of lightning's transformative power and the remains of a mouse, both contributing to the soil in different parts of the world.

To fully understand the explanations presented so far, we must shift our focus back to ourselves.

During the software creation process, you will notice many elements interacting, resulting in a unified whole. Avoid interfering with any of these processes, as they collectively complement one another. In the end, if we meet the conditions for transformation, the chosen path is indeed right.

As a developer, to contribute meaningfully to transformation and change, it is essential to avoid quickly reacting to the sudden, unpredictable shifts that occur in life. When faced with losing something precious, a new algorithm of values naturally emerges. It is in those moments that distant lightning brings forth a new reality. It is important to acknowledge that these realities manifest within our minds.

Experiencing loss in life signifies not only the beginning of new paths, but the creation of fresh lightning. Never forget that everything we lose will be replaced by something new.

> Exercise_
>
> Keep listening to natural lightning happening in real time, and always be prepared for the new.
>
> Nitrogen, an essential element for life, sparks lightning in our brains as well. When nitrogen activates our neurons, it initiates an electric surge in our minds, similar to its role in nature. Each subsequent spark enables the retrieval of pre-existing knowledge within our minds.
>
> Keep your ears and eyes open. Immerse yourself in the symphony of nature and strive to perceive its realities through deep observation.

Authentic sounds have the power to generate energy and stimulate the nitrogen in the human brain. This study area will soon become a significant focus within the scientific community.

No electronic device can accurately reproduce the genuine sound of a lightning strike. Screens and movies adapt all sound effects to a biomechanical system for our experience. Thus, they do not accurately represent the actual experience. This supports the notion that pioneers lead social lives rather than asocial ones.

In simpler terms, people currently labeled as "asocial" due to their solitary living in rural areas are, in truth, social because they engage directly with the natural environment and sounds of nature. Conversely, those labeled "social" are not truly social if they lead unnatural lives. These insights are not applicable to current social media platforms; thus, it is crucial to consider these realities while considering the disruptive elements in nature.

A developer should be able to disrupt some of the values within their programming code. When a developer disrupts these values and attempts different methods, similar outcomes may still arise.

> Exercise_

Frequently, destruction leads to the disappearance of an element, thus requiring lightning to introduce a new element and fill the gap.

Attempt to rewrite a single line of code as ten lines. By repeating that process, you may uncover the possibility of adding an extra eleventh line or even additional lines nested within it. If you desire sparking lightning in your brain, try deconstructing software you have previously created, approaching it as a playful activity. Remove certain components and observe the impact of those changes. As you can imagine, these mental lightning strikes can make you feel refreshed, much like the beginning of a fresh morning.

Chapter 2: /... /... / Life and Work: Balancing with the Algorithm

Now, let's shift our focus to the practical aspect of the subject under consideration. Ultimately, the most crucial action following this confusion is to avoid its continuation. Understanding the factors that can resolve the confusion between meaning and consciousness empowers us to make informed decisions, understand their consequences, and apply them, not only as developers, but also as social entities.

Presently, the most influential individuals in the world are recognized as "those who know what they are doing" due to their deep understanding of their actions. However, their path to success results from their adherence to principles in nature. Those individuals naturally employ three strategies that enable them to resolve confusion effectively while living within this natural state, they are eliminating inefficiency, avoiding instability and removing obstacles. We can characterize these individuals as "individuals with intuitive talent." Perhaps their accumulation of life experiences has guided them to this point.

However, anything in nature can repeat. The practice of these repetitions and their subsequent simplification hold great significance. Nature carries out its repetitions through a simple yet remarkable algorithm. In this context, it is crucial to identify the three factors that cause confusion between consciousness and meaning. Afterward, we will go over the basic exercises and clarify the underlying algorithm. This algorithm will reveal the harmonious balance between software developers and their managers. Keep in mind, however, that if team members make the same errors and mistakes repeatedly, it indicates a flaw in your approach. However, the overall balance remains unaffected.

In summary, three main factors contribute to the confusion between consciousness and meaning. These factors include inefficiency, instability, and obstacles to transformation. We will examine each factor separately, discussing approaches to eliminate them.

By incorporating these approaches into our lives, we avoid imposing rigid rules on ourselves and instead nurture a sense of joy in our pursuits. Upon reflection on our joyful experiences, we will realize that we have already rectified those functions. These rectified functions will typically have a positive ripple effect on other components within the algorithm. As one team member improves, it inspires the progress of at least two others. However, if an individual within the group stumbles, it can disrupt the overall balance, posing challenges for the entire team.

< / 1_Eliminating Inefficiency >

The hesitation that arises from a lack of productivity gradually leads to inefficiency. This outcome occurs when a person creates a clear separation between their professional and personal lives. However, our profession should naturally bring us joy and fulfillment. If someone views their profession as a burden rather than a pleasure, they may experience unproductiveness.

Our teachers often advise us to choose a profession that brings us joy, reminding us that those who hesitantly engage in their chosen profession are more likely to experience unproductivity. Therefore, even during leisure periods, try to maintain a connection with your work by reflecting on at least one topic related to your profession. Remember, thinking is a natural process. An individual may declare, "This is my day off. I will avoid thinking about work." However, when such a mindset becomes widespread, it signifies the start of a path toward being inefficient. The interconnection between our professional and personal lives makes it impossible to fully separate them. Life is a holistic experience; it is not divided into distinct professional and personal domains. Despite various rules, life continues as a unified whole. Avoid separating your profession from your personal life.

When an individual tries to avoid unproductivity, their brain experiences high levels of stress, which could lead to anxiety disorders. Indeed, it is one of the most urgent mental health issues of the present time and the foreseeable future.

I am not suggesting that you fully engage in work during your time off. Instead, it is important to be mindful and avoid adopting a narrow mindset that is limited to either "vacation only" or "work only." Your mindset should consistently align with your desired focus, as within this domain, the power of thought holds great importance and serves as a defining characteristic.

If you command your mind to not think about a particular subject, then later request that same thought, it will get confused. "Should I focus on bringing the information to the forefront or keeping it in the background?" As the brain struggles with this dilemma, the information wanders aimlessly. This wandering represents inefficiency. To produce something successfully, it is essential to have readily accessible information. Making sure information is easily available requires regular engagement with it.

Imagine yourself at home, cleaning up your room and organizing all your belongings in their proper places. Eventually, you may feel a sudden desire to use the items on your desk. Whether this desire brings you joy or remains a necessity, creating and maintaining such an organized space will protect you from various stages of unproductivity.

</ 2_Avoiding Instability >

The second factor, instability, poses a significant obstacle to society, as it relates to the failure to achieve a balance between personal life and profession. During the implementation of this mental process, we can observe the most crucial step to address this.

Strive to maintain a balance between your personal life and profession. Remember that your work is a component of your life, and your life encompasses more than just your work; they cannot be equivalent.

Life involves interacting with others, engaging in different social settings, maintaining a home, creating a personal space, enjoying recreational activities, going on vacations, and many other pursuits.

Avoid dividing the processes that make up your entire existence. Trying to categorize those aspects will only lead to a struggle to maintain a balance among them, ultimately creating instability.

Nature achieves balance naturally.

</ 3_Removing Obstacles >

The third factor relates to obstacles in transformation. This includes resolving the confusion between consciousness and meaning, creating new codes, and transforming them into something else.

To prevent obstacles in transformation and promote natural completion, observe when necessary, be a mere observer, and simply watch. Eventually, you will feel the desire to help one of the elements you have been observing.

Take a seat on a park bench and observe the actions of others. Spend some time reflecting on the origins, destinations, professional lives, and emotional states of others. Then, shift your attention toward individuals experiencing homelessness and try to understand and empathize with their emotional experiences. Observe the actions of a child playing in the park and try to comprehend the reasons behind any conflicts or disagreements that may arise during their playtime.

Examine a stray dog's behavior and try to understand its reasons for scanning the surroundings. Is it feeling hungry or thirsty? Which appeals more to its taste: bones or sausages? After contemplating, go to the nearest market and buy one of those options. Avoid purchasing both, as this exercise will serve as an evaluative measure of your decision-making. Keep in mind that when you give the chosen food to the stray dog, you are engaging in communication with it, and nature will respond with mutual grace. This interaction will help remove barriers and facilitate your personal transformation.

In life, both humans and animals serve as metaphors of transformation, reflecting their emotions. This is how history unfolds.

</ The Algorithm >

Now, let's examine the algorithm of the three key problems resulting from these practices. What would be the consequences of constantly encountering the obstacles of unproductivity, instability, and transformation?

>_ ***Inefficient + Instable***

When a software developer experiences both inefficiency and instability in their life, it could indicate significant problems in their family or intimate relationships. Such a developer is highly unlikely to achieve success unless they thoroughly address this issue.

As an analytical comparison, the algorithm of these three aspects holds great importance for human resources. People caught in this situation gradually lose their decision-making ability, making them unable to make the right choices when necessary. Often, you may hear these people say, "I had thought about that before!"

>_ ***Instable + Obstacles***

When a person faces both instability and obstacles at the same time, it is highly likely that they are dealing with issues related to circulation. If there is a slowdown in blood flow or other circulatory issues, the individual's competence in both cases is adversely affected.

To restore competence, it is essential to undergo testing for any potential circulatory system disorders. Seek medical advice and opt for natural methods to encourage recovery. Engaging in physical activity or sports may not be enough, as circulatory issues could result from pressure in certain areas of your body. Currently, circulatory issues make up 28% of health disorders and circulatory disruptions cause instability and hinder transformation.

Developers affected by such issues are unlikely to achieve desired outcomes fully and within the intended timeframes. Surprisingly, circulatory issues can be attributed to 90% of errors made by skilled individuals who are unable to meet deadlines. While mental factors do contribute, they are beyond the scope of this book.

>_Inefficiency + Obstacles

If a person concurrently faces inefficiency and transformation obstacles, this indicates an issue within their musculoskeletal system.

The human body's ability to perform tasks efficiently is driven by its muscular system. Insufficient or improper development of the ligaments can lead to unexpected bodily discomfort in the long run. The pursuit of success at work intensifies the motivation to overcome these discomforts. A developer unable to overcome this issue will either give in to it or resort to dishonesty.

If all three elements (unproductivity, instability, and obstacles to transformation) coincide, experts recommend the developer take a break from coding and completely rejuvenate themselves. Your mind may be the source of your abilities, yet it is crucial to remember that your mind is a part of your body. Never ignore the fact that changes in your body can impact every aspect of your life.

Chapter 3: / Continuous Production

Every result achieved by developers initiates a new production process. The end user, interacting with your application or program, benefits from the product and continues to produce new aspects. In essence, all production phases start again when a product launches. However, your actions triggered this ongoing process. You lay the foundation for new aspects. The ultimate goal for both you and the end user is to continue this process.

Recognizing that production is an ongoing process, you must keep up a consistent flow of your own work and keep producing. Examining the history of production in the past century, it becomes evident that both *vertical* and *horizontal development* play vital roles in sustaining continuous production.

When purchasing software developed by a developer, the buyer expects the product to include three essential components.

1. The first component is competence. The software we use gives us control, a power we try to use to achieve our goals. Meeting the needs of an organization creates the necessity to activate this competence.

2. The second is operation, which includes features that save time and money, leading to improved efficiency. Concurrently, we expect the team to achieve excellent outcomes while working together on a project, each focusing on their own areas of competence. As those processes move forward, we encounter new obstacles. We find ourselves in need of a software that consistently resolves our issues with updated versions and programs. This need is primarily for the user and secondary for us. While addressing our needs, we unintentionally trigger horizontal development. This means our development will enhance the competence of our software.

3. The third component for the developer is self-development. A development confined solely to the realm of software proves insufficient. While we dedicate significant time to crafting software, the end user may

integrate the program into various aspects of their life—from daily routines and social interactions to application within kitchen or garage environments—accomplished through simple clicks.

If the developer responsible for providing the end user with competence finds themselves disconnected from such a lifestyle, the likelihood of achieving genuine results from their work decreases significantly.

For a program to function correctly, diligent work is essential. You must have a strong passion and determination that allows you to add value, thus continuously improving the elements involved. Indeed, patience is crucial to becoming a developer. Patience, in essence, signifies the ability to sustain one's existence in the face of all adversities encountered along the journey.

Development exists in two forms: vertical and horizontal.

< / 1_ Vertical Development >

For a software developer, vertical development is a progression from personal development to work and finally to the practice of patience.

< / 2_ Horizontal Development >

The horizontal development involves using the achievements gained from vertical development, triggering problem-solving mechanisms through competence, operation, and patience.

Comprehending your true vertical and horizontal state is vital when building a team, assessing your role within the team, or carrying out your work responsibilities. Start anew by noting the principles of development, work, and patience. Afterward, assess your competence in the stages of horizontal development, how active you are with your competence, and your ability to solve problems using your competence and activity.

Chapter 3: /... /... / Convex, Concave, and Harmonic Developers

Recognizing your position within the graphics later in this chapter can provide insights into your developmental progress. At first, you might notice a rapid increase in your growth curve, but as you approach the point where patience in vertical development and problem-solving in horizontal development align, your progress will decelerate, resulting in a less noticeable curve.

You should not perceive the x and y axes as always moving together. It is essential to comprehend the meaning behind these curves, so set aside time to understand them.

Let's analyze this developmental graphic in its entirety. Imagine the elements of vertical and horizontal development placed on an xy plane. Start at the origin (0,0) and position "development," "work," and "patience" along the x-axis, representing vertical development. Likewise, position "competence," "operation," and "problem-solving" along the y-axis, representing horizontal development.

Within the software world, except for extreme points, three distinct types exist.

</ 1_ Convex Developer >

In this first type, development comes before competence in the progression.

These individuals devote a lot of time to reading and learning new information. At first, it may confuse others as to why they have difficulty making decisions effectively. Their decision-making mechanisms are not immediately activated; instead, they read and analyze beforehand. They often request "extra time to work." These individuals also give themselves this time and maintain a sense of ease. As they develop their skills to a satisfactory level, we can observe an increase in their competence. Their operational state emerges once they exhibit patience and reach full maturity.

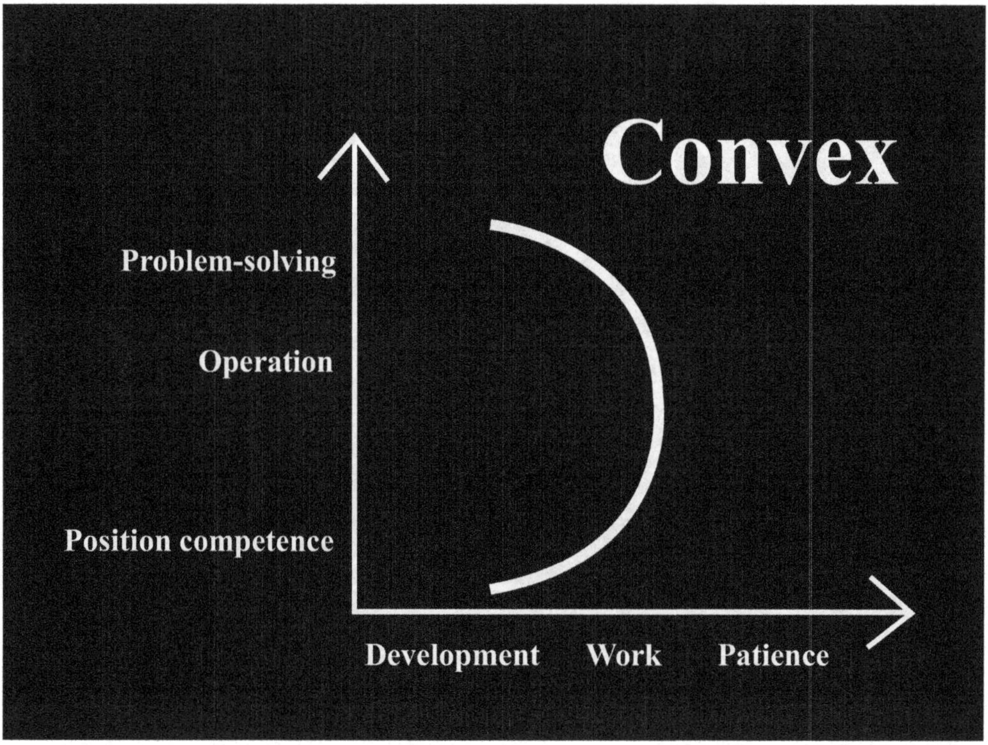

In vertical development, problem-solving advances along a convex line, like the one depicted in the preceding figure. These individuals are software developers. Their strong focus on personal development slows down their operation and competence, but they are also the ones who originate and quickly adapt to new ideas, pushing them forward in the software world. Currently, they are known as "prodigies."

</ 2_ Concave Developer >

The second type aims to attain competence quickly. These individuals are deeply involved in decision-making processes and have a strong desire for development. As a result, they maintain their social and professional lives along a concave curve.

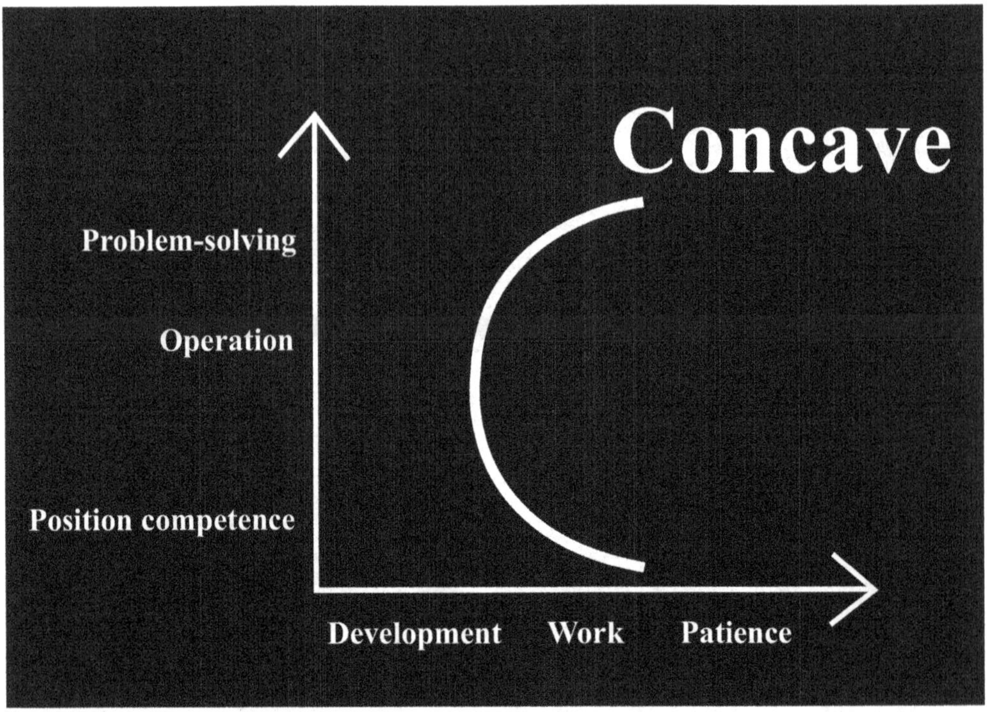

Upon reaching the stage of operation, they become considerably industrious. As they move into problem-solving, their concave curve inclines toward patience. Typically, these individuals display a "can-do" attitude and make swift decisions. Despite their mistakes, they remain undeterred, becoming increasingly focused on their goals and quickly self-motivating. They are the operators, quickly implementing actions and taking leadership rather than falling behind. Upon receiving a task, they use all the necessary actions to complete it.

To software developers, this approach may initially seem somewhat chaotic. However, both types follow the same direction within the graphic.

</ 3_ Harmonic Developer >

The third type represents a more harmonious progression. These individuals embrace both development and competence before taking a momentary pause. They assess the potential of their operational activity and try to envision its future trajectory. At this point, they observe life once more before moving on to another level. Upon comparing these processes, it becomes clear that they constantly break down and reemerge in various forms.

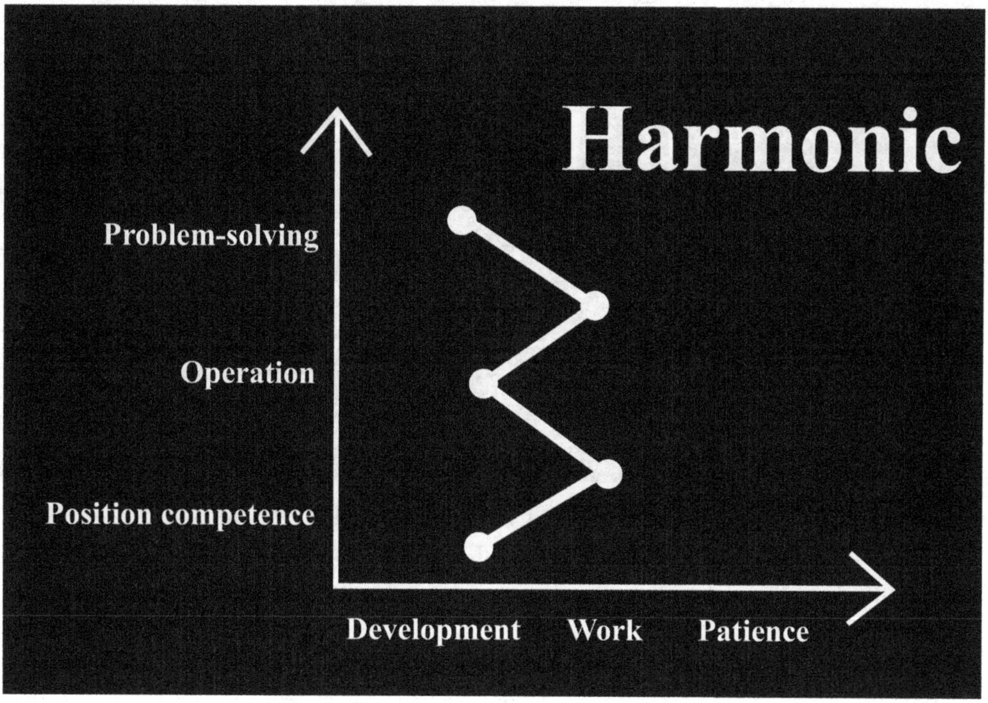

As a result, a figure resembling a sidewards seagull appears in the developmental graph.

</ 3_Coffee Bean Developer >

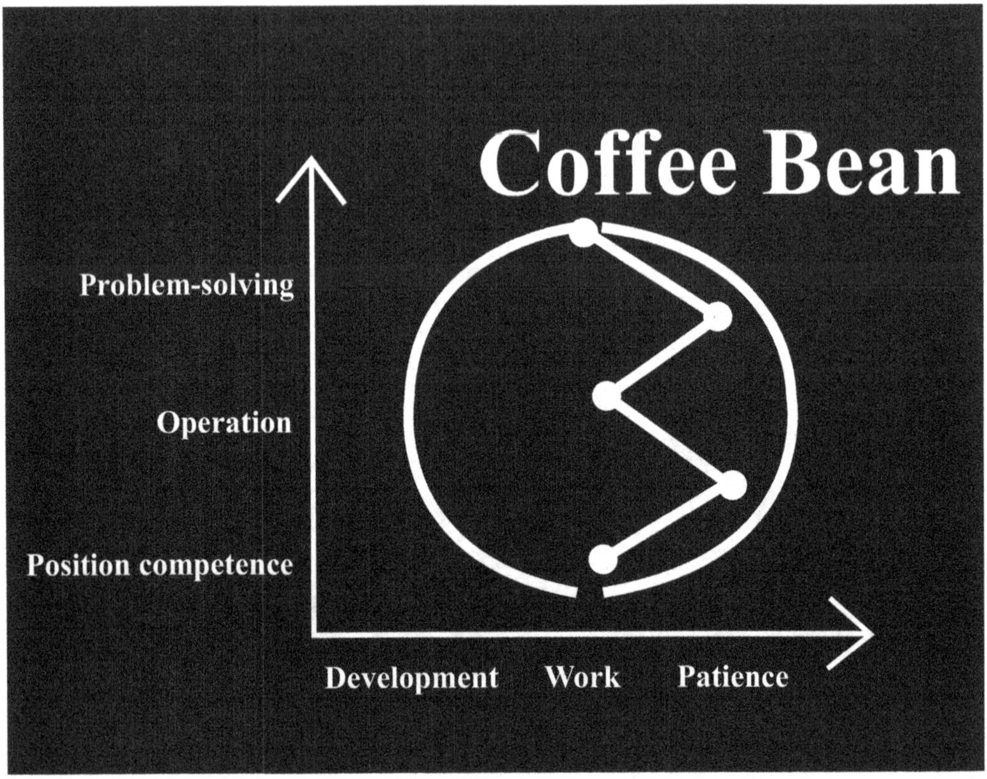

Looking at the graph from a wider perspective, you can see that the concave and convex structures intersect at a point. This space, resembling a coffee bean, represents leaders.

Indeed, a coffee bean can tell us a lot about the software world. Their starting and ending points share similarities. The depth of the line in the middle gives the bean its unique shape, smell, taste, and color. The roasting process is just as important as the soil quality where the bean grew. Both factors connect to the central line and the two curves that surround it. Therefore, when considering individuals for your team, it is essential to consider their positions in this graphic. If the resulting figure resembles a coffee bean, you are certainly on the right track. Be confident that others will surely appreciate the metaphorical "coffee" brewed from this particular bean.

Chapter 3: /... /... /... / Enhancing Growth through Constructive Processes

Now, it is necessary to explore how to enhance the shared qualities of individuals within your "coffee bean team." What activities might help to stabilize a solid level of development, work, and patience? Keep in mind that each individual may need varying amounts of time to achieve similar levels of development.

The most crucial factor of importance and value lies in the effort put forth by each team member, rather than the time individually invested. The effort put forth during this journey will lead to better decision-making, quicker actions, and superior results. To enhance development, engaging in activities like assembling Lego sets and solving puzzles can be beneficial. Building with blocks and creating mockups significantly impacts your transformation and completion, helping align them with the points in your brain's algorithms. Taking part in such activities helps reduce your stress and teaches you how to work on your design.

Once you have become skilled in the necessary steps, your improved motor functions will help you achieve a more productive state. Now, imagine someone talented at painting, perhaps even an accomplished painter. However, if they were asked to write something in calligraphy, they would require some time to accomplish this task.

One might assume that an individual skilled in painting would also be proficient in calligraphy. However, such an assumption might not be accurate. In this situation, the brain requires a new construct to improve its existing skills. It's not that the painter must spend time learning calligraphy as a new skill; instead, they must spend time practicing calligraphy to improve their motor skills.

A developer should always maintain active and open phases of development in constructive processes. When these phases end, the individual no longer enjoys the production process, leading to a shift. This shift leads to a mindset of obligation driven by financial necessity, ultimately hindering personal growth and development. To avoid such situations, one should not hesitate to invest in tools like Legos or 3D puzzles. Engaging with these tools patiently will stimulate the brain's developmental functions, promoting constructive progress.

The next stage concerns the conditions for work. After reaching a particular level of development, the brain necessitates action and movement. To achieve this, the resources must be readily available.

An important aspect to consider is that stress levels increase unexpectedly when a person feels compelled to work to meet their personal needs. This prevents further progress and leads to a greater number of errors. To unlock the proper working mechanism, engaging with the lives of other human beings and animals is vital.

While observing others, our minds naturally ask, "How can I contribute to their well-being?" This line of thinking triggers us to act. Therefore, if you desire a self-motivating mindset, offer your help to individuals and animals in need.

The widespread food shortage affecting humanity deeply saddens us all. Offering help and support to those in need not only brings happiness within us, but propels our personal and professional growth.

If you are considering pet adoption, cats, dogs, and birds are popular choices. However, before bringing an animal into your life, it is crucial to assess your personality and lifestyle carefully. When choosing the ideal companion, it is essential to consider its potential impact on your life rather than solely focusing on personal preferences. Maybe you may have a strong desire to care for a pet, but you want something requiring less maintenance. So, even though you might initially consider a bird or fish, a cat or dog could be a more suitable companion.

Communicating with an animal that mainly flies, like a bird, might be more challenging. If you are an individual who does not need constant interaction with others, birds might be a suitable choice as pets. On the other hand, if you enjoy social interactions and frequent engagement, bird companions may not provide the most suitable companionship.

Fish, staying in one place and relying on their caretakers for food, make suitable companions for individuals who prefer a home-based lifestyle. However, if you enjoy the outdoors, spend time in parks, or frequent cafes like Starbucks, keeping fish as pets might cause some stress.

While both cats and dogs can evoke similar emotional responses, cats have a more home-based nature, while dogs show a stronger preference for outdoor activities. As a result, taking care of a dog may encourage a stronger bond with nature, which could interfere with your work; therefore, in such situations, a cat might be a more suitable option. However, if you prefer performing tasks in natural environments, choosing a dog would be a better match for you.

All the steps and processes mentioned before require a hardworking, developed person to show greater patience. Indeed, events may not always happen as expected, and each step comes with new challenges. Patience is the most important virtue to solve these situations effectively. If you are naturally impatient and hasty, be aware that you will initially face challenges in solving problems in your life.

> Exercise_

A beneficial exercise to enhance patience is growing a plant not well-suited to your local environment. Obtain a plant from another country or climate and try to create the best conditions for its growth. That may require building a small greenhouse or terrarium inside your home. Every effort you put into nurturing the plant will help increase your patience for future endeavors.

Keep in mind that individuals who do not focus on self-improvement cannot function effectively, and those who cannot function effectively cannot cultivate patience.

Let us compare the characteristics of nature and our previous discussion regarding the potential impacts of an unnatural lifestyle. Since we are a part of nature, this comparison will help us better understand natural development processes, particularly regarding individual development.

When we observe the universe, we notice there are more vertical aspects than horizontal ones. This indicates that the universe is well-developed, continuously active, and patient—an entity capable of responding to changes at the right time. After it reaches this point, competence, operation, and problem-solving come into existence. For example, within an ecosystem, elements like forests, lakes, and wetlands are naturally designed to meet the needs of the animals that live in them.

However, when nature faces pollution, it must react with competence and effectiveness through the help of bacteria and plants. Among these are the pseudomonas bacterium, known for its ability to degrade certain pollutants, and plants like vetiver grass, renowned for its capacity to tolerate and absorb heavy metals from contaminated soil and water. Nevertheless, this process may take time for these helpful organisms to reach the affected area.

This topic is important in biology, but to summarize, many modern studies depict food chains, human life cycles, sociological structures, and similar concepts, like Maslow's hierarchy, using pyramid or tree-like structures. In a universe where vertical aspects prevail, it is clear that nothing exactly mirrors the tree algorithm, which acts as the foundation of our algorithms.

The topmost part of the pyramid/tree holds a particular value that expands downward, giving rise to the algorithm. According to that algorithm, an individual at the bottom can take specific steps and ascend to the top. However, nature does not possess such a structure.

Nature indeed showcases a dynamic and efficient structure. From a far-off perspective, it does not exhibit pyramid-like features or embody pyramidal characteristics; instead, it displays a web-like, interconnected structure. Every individual and object occupy a distinct level within a regenerative and self-improving structure that is always active.

The core of these values suggests the necessity of abandoning the tree algorithm, which would impact all mind-mapping processes essential for brainstorming. If you aim to achieve a goal, avoid trying to outline its downward expansion using the mathematics of a tree structure.

Each individual is a magnificent piece of art. Certainly, humans are the masterpieces of nature. Thus, it is not recommended to place your goal at the highest point and create a sequence of steps below it, whether for software development or any other professional undertaking. Instead, a cyclic approach might be more conducive, where each step or task is part of an interconnected web supporting and informing each other, rather than a linear progression toward the goal.

Things that are not part of nature are unnatural, and such things are not well-suited to humans, who are nature's masterpieces. Imagine yourself as a building. If there is a problem in one area of the structure, the other floors, representing your various qualities, will make up for it. This system ensures that minor issues do not cause the entire building to collapse. As long as life continues, the building remains standing. A flaw on one floor will not affect the entire structure. The positive qualities within the building help eliminate the negative aspects.

Do not be misled by the brain being at the topmost position, like the tip of the pyramid. All organs can make independent decisions and compare them with the input from the brain.

To illustrate this point clearly, consider the building example we just discussed. Just as a flaw in one floor does not threaten the entire structure, an issue in one organism does not automatically dictate the overall health of all organisms. Organs like the heart, liver, or intestines possess internal mechanisms that operate independently of the brain to control, mend, and handle problems. For instance, if there is a malfunction in the liver, it has its internal responses to address the issue, regardless of the brain. As a result, these structures show that ongoing problems are not solely linked to your mind, but also to your organs and the overall systems.

When a software developer fails to achieve their goal or complete a project, the most common mistake is doubting their mind or focusing on their perceived incompetence. However, the truth is that we cannot solely attribute the problem to our brains. Under such circumstances, it could be essential to examine the brain's neural system, as well as the transmission of signals through the spinal cord. In such cases, massage therapy could be helpful.

As mentioned earlier, problems related to circulation can manifest as kidney discomfort, fatty liver syndrome, heartburn, or gas. Each of these symptoms can significantly affect our decision-making mechanisms.

More research on this subject is crucial. For instance, joint pain may not only result from a sedentary lifestyle, but could also be linked to circulatory issues. Basic walking may not be enough to improve circulatory issues for individuals whose jobs mainly involve sitting.

On the way to becoming a successful software developer, our bodies must function like a harmonic structure. Once all parts work in harmony, the developer should focus on improving their competence in their chosen area. This competence in development pertains to understanding what they're doing.

If a developer struggles to grasp what they are doing and cannot envision why they are doing it and for whom, they will not achieve competence in their coding language.

Think of yourself as the end user of your application. If you are not the first to use it, or you cannot empathize with the users' experience while creating it, you will inevitably face challenges in becoming competent.

For the effectiveness of your software, understanding the tools and methods you are using is essential. Achieving this level of understanding requires experience from previous versions or similar projects. Avoid dismissing such experiences as "just basic software" and postponing the learning process. It is precisely through these experiences that one cultivates as a developer. Remember, the simplest applications continue to be used worldwide. For instance, consider the basic text editor applications like Notepad on Windows or TextEdit on macOS. These programs, despite their simplicity, remain essential tools for a wide range of users. Taking another look at and attempting to rewrite these applications can be a great exercise. Every mistake you encounter during the process will reveal new paths for you, providing opportunities for progress and improvement.

Start with reprogramming the code of an old but still in-use application with no delay. Avoid adding any active elements to the code initially. If you wait to do so in the next step, it lets you observe the activity and identify what stays inactive.

Remember, it is not the act of someone pointing out your mistakes that is important. Instead, it is the intentional making of those mistakes, followed by careful thought and correction, that allows one to learn the concept of errors truly. If you do not learn the concept of error, it triggers a rejection response in your brain. As a result, you might unknowingly repeat the same mistake. These error-focused practices clarify the principle of problem-solving.

Creating code naturally requires fulfilling three essential requirements.

- Knowing what you are doing.
- Knowing the tools or methods used.
- Knowing what you are changing.

Once you have a thorough understanding of these requirements, you will be able to see what you have completed. Once you comprehend what you have completed and changed, along with your contributions through the steps, your journey to becoming a real developer will begin.

Considering all the factors mentioned, we will encounter something new. You may not notice it immediately, but your code establishes a new balance between usability and accuracy. Alternatively, it claims to find a new balance. The closer this claim aligns with reality, the more essential this program becomes for individuals and end users.

For instance, consider the historical tool known as the abacus. In the past, people used the abacus to perform mathematical calculations. These individuals faced two key challenges. Upon examining these challenges, we discover what balance requires.

1- The first problem was calculation errors. A single mistaken bead could lead to incorrect results, and many such errors were possible. Research in Egypt has shown that using abacuses in accounting led to numerous errors, necessitating the appointment of a supplementary individual to monitor the process.

2- The second problem related to time. As the volume of calculations increased, the abacus needed adjustments, leading to significant time losses. To counteract those time losses and errors, a new tool was necessary. At that point, the modern calculator, representing the latest technological evolution of the abacus, became integrated into our lives.

We need a tool that does not make mistakes or waste our time and can quickly solve our problems. This constitutes a proposition. The design of this proposition takes shape in the response to people's needs. Each developer also contributes to fulfilling these needs.

Let's explore the binary relationship between the abacus and the calculator. As we learn more about their functionalities, we also gain insights into reducing errors and saving time from a clearer perspective. In the past, when using an abacus to calculate the total number of pears, one had to use both their mind and hands actively. The exchange of information was based on the quantity of materials. Writing the number of pears was enough for calculation, making the material turn into information.

Nowadays, you do not need to have these things in hand because software developers turn them into digital data.

As materials transformed into information, machines replaced the human mind. Even though humans created these machines, they are now the main storage of data and results. Hence, a new balance emerged between material and information, as well as between the human mind and machines. This balance digitized the matter and mechanized the mind.

Currently, every program we create converts the physical dimension into a specific binary coding system of ones and zeros. The data we input undergoes immediate processing, providing value instantly. This principle is applicable in our lives as well.

As developers advance information and machines in contrast to materials and the mind, they simultaneously cultivate a new balance within themselves. Every individual strives to understand their position within this new balance. This is because the present moment within the dynamic flow of time represents the moment we seek to shape our desires and passions for the future.

We can term this prevailing circumstance—the ongoing technological and informational evolution—as "conditions." Within the construct of these conditions, we are under the belief that we can establish a new balance in accordance with our individual capacities.

Currently, we are working to establish the future and capacity that will confront information and technology within those conditions. We are doing this as a trade-off for the materials and mind we previously discussed. Thus, we imagine the future today, seeking a capacity that exceeds current conditions. It is crucial to recognize that today is, indeed, the future, and these conditions are shaped by our capacity. At this point, it becomes essential to rewrite the elements of balance.

> Exercise_

Grab a sheet of paper and a pen. Write down this idea: everything that exists today is actually the future, and all the current conditions reflect our capacity. Once you embrace this idea, you can confidently move toward the future, making the most of the capacity that your present circumstances provide. By embracing the future of today, we inevitably leave something behind: yesterday.

We cannot fully experience the present if we cannot let go of our connections to the past. Every moment not focused on the present indicates a lack of potential for the future.

When the future finally arrives, you may find yourself not having the needed capacity.

It is crucial to let go of the past and realize that the current conditions are sufficient for building a better future. As an individual, it is crucial to recognize that you are the one who decides and shapes your own future. To create a future, especially as a developer responsible for creating this information, you must maintain a balance within your field and social relations. To achieve this balance, it is essential to let go of the past.

Not letting go of the past can lead to your brain and body becoming exhausted during the night. In such a case, even with the help of morning coffee, getting yourself together can still be challenging. Therefore, it is extremely important to fully let go of the past and focus on the future. An unresolved past in one's mind continues to create obstacles in the present, preventing the use of the capacity to move forward.

> Exercise_

To address this problem, spend five minutes before going to bed doing this exercise. Sit on your bed and think about those who have lied to, betrayed, or hurt you. If you can forgive them, do it. Forgiveness is not mandatory. If you see the potential for resolving the matter the next day, reassure yourself of your intention to address it. On the other hand, if you think the problem will persist, do not hesitate to let go of the past. Adopt this mindset and include this habit in your daily routine. By doing so, your brain will act as the examiner of your past.

From now on, the day you experience will truly be the present. This present day represents your true future. Moreover, your dreams serve as a platform to encounter realities. If an individual cannot let go of the past and concentrate on the future, expecting their contribution to the future is unrealistic. Remember, a developer's primary responsibility is to work toward a better future.

Is this five-minute exercise of letting go of the past and preparing for the next day enough? Absolutely not. Click here to enter text.Every exercise must be accompanied by an action before and after. Otherwise, these exercises are bound to be ineffective.

Chapter 5: / Living in Harmony with Nature: Nature, Mind and Body

Many self-improvement books talk about journaling, time management techniwues, positive affirmations, manifesting but they often fail to explain how to incorporate them into your daily life. As a result, the provided knowledge often falls short in practicality. While the ideas presented in these books may be valuable, and some may align with the exercises mentioned here, it is crucial for the practices to fit seamlessly into your current daily routines.

For instance, imagine you were diagnosed with cancer. When you receive the diagnosis, your doctor prescribes medications and advises you on your diet, sleep routine, building positive relationships, keeping away from negative influences, and maintaining a positive mindset. Those recommendations are crucial because the prescribed medications will interact with and influence various aspects of your life. In fact, many individuals have overcome cancer by adopting a positive mindset, healthy eating habits, and making precise lifestyle modifications.

This example emphasizes the importance of additional practices alongside treatment plans, as they have a significant impact on the primary healing process. With this understanding, let's delve into an essential aspect of personal development: fostering a connection with nature.

Spend time in nature when you feel bored or need inspiration. Engaging with nature, especially during happy moments, can bring significant benefits, and over time, these practices mentioned will become a regular part of your routine.

> Exercise_

For instance, try observing bees in their natural environment to truly let go of the past. This action reflects the significance of diligence, concentration, and progress toward future goals despite previous obstacles. Observing bees means more than just watching videos of bees; it involves experiencing their behavior firsthand.

Consider setting up a small beehive in a natural environment. The happiness you will find from them producing honey and immersing yourself in nature is immeasurable. Watch the bees diligently building their honeycombs and increasing their honey reserves. You will see their unwavering dedication to securing their hive's future as they dutifully serve their queen.

As you enjoy the honey, the result of their hard work, you will be drawn to understand its source: the flowers. Each flower and its pollen play a role in determining the honey's unique taste. When you recognize this, you will find yourself eager to learn about the plants in your garden. You will gather some flowers, take pictures of others, and dive deeper into studying them to understand their true nature.

Study their properties, benefits, and uses. If bees are attracted to flowers, it indicates they have medicinal properties. For instance, flowers like chamomile and lemon balm may help improve sleep quality and increase energy upon waking up. If you make tea using chamomile and lemon balm and drink it one to two hours before sleep, you may find that it improves your ability to reflect on your past and current situation.

As you observe the bees tirelessly seeking these flowers, you will gain a deeper understanding of determination and patience. Enjoying the honey produced in your garden, you will appreciate the space you share with the bees. This experience will go beyond mere enjoyment.

In harmony with nature, you will recognize your efforts to create natural balances within the artificial world of software development. This understanding will help you attain a well-balanced state in your social, intellectual, and professional life.

Research from different parts of the world indicates that those who exhibit the most balanced traits both physically and mentally often live in rural areas, where they engage in activities like farming, fishing, or taking care of animals. This trend arises because these individuals do not impose themselves on nature; instead, they benefit from its complementary elements. We could argue that developers engaged in agriculture and livestock are the best tool-solution partners when applying this concept to software development.

Engaging in agricultural activities can bring significant benefits. Gaining knowledge about agriculture not only helps you connect with nature but also opens new opportunities for developing innovative agricultural software in the future. Through such experiences, you can understand the importance of connecting with the organic balance of nature to achieve personal balance.

Chapter 6: / The Present as Future: Moving Beyond the Past

While we have thoroughly discussed the concept of nature, addressing potential obstacles to change is also crucial. Otherwise, it might seem that our discussion is overly idealistic, giving the impression that nature always leads to positive results. Indeed, some aspects of nature may seem strange or lack a clear purpose. We might perceive these seemingly meaningless elements as obstacles when they come together. In such cases, it is wise to step back temporarily and look at the entire picture as a whole, becoming a neutral observer.

Taking on the role of an observer is not as unpleasant as it may seem at first. A common issue in modern society is that many people have lost the ability to watch and observe patiently. In certain situations, such as during team meetings where quick decisions are required, or when facing tight project deadlines, we tend to make hasty decisions and act impulsively. Software developers, who are constantly exposed to quickly changing visuals and data, immersed in coding, and struggling with the lack of time for other pursuits, also face this issue.

When we examine global production factors, we notice their impact on the natural balance. Disruptions to this balance can create complex challenges that aren't always obvious.

Consider rainfall. Clouds need to form before it rains, and when the right conditions align, wind collects these clouds, resulting in rainfall. In regions with low precipitation, the interaction between rainfall and dry soil is complex. Contrary to the assumption that dry soil quickly absorbs water, it can actually be hydrophobic, leading to water runoff rather than absorption.

While rapid soil absorption of water is not a universal phenomenon, nature has developed efficient water storage methods in certain ecosystems. For example, cacti in deserts store water to provide hydration for themselves and other desert dwellers. In deep valleys and deserts, oases, often formed by underground water sources, are crucial for the survival of many species. These examples illustrate how nature adapts to varying water availability, but the generalization about soil absorption as a strategic water reserve is an oversimplification.

In the context of natural conditions, the obstacles in nature do not truly hinder progress but rather serve as balancing elements for other processes. Our focus should then shift from the perceived obstacles themselves to the growth they stimulate. By adopting this perspective, we can grasp the hidden messages nature conveys.

It is important to remember that all significant software projects have faced serious mistakes. Well-known platforms like Facebook, Google, and YouTube have made substantial mistakes during and after programming. For example, Facebook has faced privacy concerns and data breaches, notably the Cambridge Analytica scandal, which involved the misuse of user data. Google encountered criticism for its involvement in Project Maven, a Pentagon project using AI for drone targeting, raising ethical concerns. Additionally, Google+ had security flaws exposing user data, leading to its shutdown. YouTube has struggled with algorithmic issues, inadvertently promoting harmful content, and challenges in content moderation. Nature opposes any system that disrupts its balance and uses various methods to counteract it. To put it more accurately, nature provides a counterbalancing force that supports establishing a new system.

A pushing force does not always provide the support; sometimes, a pulling force can equally serve a supportive function. Therefore, you should not obsess over the origin of errors. Instead, you should emphasize the results of their occurrence. When an issue resolves, the primary focus should not be the solution itself, but the new development that it established.

Chapter 6: /... / Learning from Mistakes: Discovering Novelty

Every mistake generates something new. The elements you change or regenerate in a software project are not your own work of art. Despite your original intentions, the artwork inevitably transforms into how it should be. One could argue that those processes are interconnected and may result in a sense of overwhelming complexity. At this point, realizing that the main problem stems from seeking perfection is important.

Chasing perfection is unnecessary because trying to reach an unattainable goal often hinders progress. It is not about how you can fix mistakes—they will inevitably resolve somehow, given your competence and operational skills required. Ultimately, you will overcome the problem, whether it is tomorrow or with the help of others. The most important thing here is the knowledge, experience, and growth you gain after solving the problem. Consequently, another approach presents itself. In software development, one should not seek perfection. Instead, deliberately make errors while coding.

> Exercise_

Reward a team member who has intentionally made an error. Identify the individual who has produced the best error or left behind the most fascinating bug. You might even think about organizing a competition or starting a systematic challenge among the team: the code with the most remarkable error.

These mistakes will inevitably reveal unexpected possibilities. Each mistake signifies a shift in balance, and each of these shifts holds a new meaning. This meaning can be so profound that it allows the addition of a new element to your code, much like a detective film where the perpetrator's identity unfolds through an unexpected twist.

The software developers of the future will move beyond just writing code; they will become architects of digital ecosystems, blending creativity with technical skill and go beyond the roles of merely protecting our cybersecurity. Developers skilled in integrating meaningful errors within robust software will have a greater impact, thus promoting the progress of advanced, futuristic, and next-gen technologies.

Go ahead. Make mistakes and observe them.

Chapter 7: / The Human Algorithm

This chapter aims to partially explain the framework of the concept of spirit. Simultaneously, we will thoroughly explore the similarities between our body, spirit, and neural system.

Let's begin with two questions.

1. Which parts of our body correspond to the elements mentioned in Chapter 6?

2. If we thought of ourselves as a computer, what real values would we encounter?

To explore this, we need to delve deeper and examine the details.

We can compare our physical body to hardware. Biological processes, like software, create a functional system with all cells working together in harmony. We can refer to this phenomenon as "mechanical." Moreover, experts often describe genetic structures as software. However, the DNA helix itself does not function like software; it is a part of the hardware. While genetic processes determine the arrangement of certain parts, they do not correspond to the dynamic movements present within the hardware.

The genetic code is an inherent part of the hardware. Our neural system houses the actual software aspect of this genetic code. When needed, our neural system demonstrates how to use the existing hardware. However, an extra element is necessary to bridge the gap between this neural software and biological hardware. In our life, we call this element experience.

Every individual may respond differently to the same situations. Therefore, it is important to recognize the existence of a unique experiential process that is separate from the software and hardware within us.

While some individuals show affection for dogs, others exhibit fear toward them. Our genetic makeup does not naturally ingrain these reactions; however, certain fears related to animals can accumulate. This introduces a new field in which AI will face significant challenges. It will need to respond differently depending on the circumstances and based on the location and time of the event.

This emotional intelligence is unique to humans and other living beings and cannot be bypassed by modifying genetic code; it shapes our perception and connects to our lives.

Let's consider another perspective. Previously, we discussed work, development, and patience. To work, one must use their hands, arms, head, eyes, fingers, and other body parts. These parts make up the hardware mechanism. The neural system, acting as the primary software of this mechanism, supports your development. Therefore, although you might receive suggestions for a more straightforward method, you may prefer certain code substructures and find working with them more comfortable.

Sometimes, you may receive suggestions to use a simpler method, but you might prefer a different one. At this point, feelings like patience, love, hate, rejection, and acceptance come up. These feelings are part of our emotional realm and we cannot solely attribute them to hormonal factors.

Research shows people with similar hormone levels may not enjoy the same experiences. We can credit the difference to unique spiritual processes. Essentially, regardless of your comprehension of the spiritual processes, they serve as the primary mechanism governing responses to structural forms. At this point, we can say that the gaps between the axons and dendrites within our neural system represent our spirit.

Contrary to the assumption from a hardware and software perspective, we have discovered that our neural system is not a continuous, singular cable. Instead, it is composed of numerous gaps, known as synapses. Our preferences reside within these synaptic gaps, resulting in our experiences differing markedly from one another. Therefore, certain tendencies may be more prevalent among some individuals than others. This makes it evident that we necessitate a balanced approach encompassing all three realms: body, spirit, and neural system.

It is important to remember that the first task is to establish a certain balance within ourselves. While the individual parts of a system may achieve balance, it does not guarantee the overall balance of the entire system. Here, the influence of others and our reactions to them arise, highlighting the need for a transformation process.

In this metaphor of balance, all stages of hardware, experience, and software work harmoniously, meeting each other's requirements.

For example, a sudden desire for a sour apple might happen, instantly triggering the perception of sourness in your taste buds. Even your sense of smell may be involved during this moment. However, our hardware does not inherently program our structure to crave a sour apple. It is here that our spirit, functioning as an experiential metaphor, comes into play, and we form the thought, "I crave a sour apple."

On the other hand, our neural system, functioning as software, reminds us of the necessary steps to procure that apple. These steps involve walking, going to the marketplace, finding the sour apples, returning home, and finally satisfying our experiential needs through our hardware activities.

Is this just a coincidence? Definitely not.

Think about it this way: imagine a woman with a painful ear because of a cold. Her body needs a particular substance found in sour apples. While her body craves this substance, saying that this demand comes only from the body would be wrong; it is also a desire of the spirit. This response to the demand activates the hardware with the tartness of the apple. Simultaneously, the software coordinates the needed qualities and amounts, ensuring the body can obtain the sour apple.

None of those preparations happen separately. Without them occurring in a harmonious process simultaneously, comprehending our desire for a sour apple would be unlikely. This system doesn't always operate in favor of positive outcomes; sometimes it serves negative purposes. Therefore, when considering the spirit, it's essential not to conflate the concepts of good and bad.

A notable characteristic of humanity is our tendency toward potentially harmful elements. People tend to underestimate evil deeds in areas where the law is weakened, and there are no control mechanisms to fulfill their immediate desires quickly. In doing so, they become inclined toward evil. This does not stem from a natural evil in humans, but from negative learning outcomes caused by their experiences.

The steps to achieve spiritual balance will be explained in detail later and it is crucial to remember that these result from the balance between hardware and software.

When creating a program, not everyone thinks about the machine it will run on or its technical aspects. Some also do not focus on the development of the software process. The most important thing for them could be the experiential value the software brings to their lives, and that value determines their preference for it.

Some software projects may take years to complete. However, if the resulting product does not connect with the end-user experience or if either the software or hardware becomes too dominant, people are unlikely to want to use it. Thus, the balance among hardware, experience, and software in a developer's life—mirroring the balance within their body, spirit, and neural system—inevitably influences the end-user's body, spirit, and neural system. This influence represents the real algorithm of the new era. When this algorithm falters, all these processes become problematic.

When the balance among the body, spirit, and neural system is disrupted—

socially, culturally, or economically—these processes become disorganized. Nevertheless, we are aware that all can still be rectified. Suppose we successfully balance ourselves—our self, body, spirit, and neural system—under certain conditions. In that case, we will achieve balance with our software and hardware, transforming us into individuals who initiate new experiences. Otherwise, failure would be an unavoidable outcome.

> The Body_

Let's start by looking at the balance within our body. While a detailed look at the body parts is not our focus, we can talk about the main issues related to development. Later, you can dive deeper into these topics and enhance your knowledge by reading more.

Understanding the primary systems of the body and how they interact can bring significant advantages to the field of software development.

The human body comprises three key systems: the circulatory, digestive, and urinary systems. The development and health of the human body depends on the effective collaboration of these three primary systems.

Development goes beyond simply growing from infancy to adolescence and eventually reaching maturity. Over time, offspring will emerge as the result of this continuous growth process.

Sexual life provides another reference point highlighting this balance. We can assess balance from various viewpoints; development and sexuality are just two aspects. However, let's first examine the three main components that create this balance.

If the circulatory, digestive, and urinary systems fail to work together harmoniously, it can disrupt both our development and sexual life. In such situations, the body's software tirelessly works to fix the hardware, which could lead to a nervous breakdown.

The human circulatory system starts with the heart and lungs. It begins with breathing and involves the transfer of essential proteins, minerals, and amino acids from our digestive system to the rest of our body. As a result, two primary components of the circulatory system need to stay balanced: the heart and the lungs. Proper breathing and ensuring smooth blood flow are vital to maintaining this balance. If someone experiences lung problems, it can disrupt the balance of their circulatory system, leading to issues in their digestive and urinary systems, too. These systems must work diligently to restore a new balance. In such situations, liver function test results might show surprisingly unusual outcomes.

Many people believe that medications for conditions like fatty liver or elevated cholesterol will bring back balance. However, without addressing the underlying circulatory problem, the body may end up in a futile state of imbalance, leading to significant weight gain or worsening of the existing condition.

Signs of primary circulatory problems often show up as fluctuations in body temperature. For example, if you experience abnormal heat or cold in your feet during the early morning hours and a doctor cannot find a solution, it may indicate a potential issue within your circulatory system. Circulatory problems typically relate to the health of veins but also depend on how well the lungs function. Factors like smoking and alcohol intake cause harmful effects on the circulatory system.

To maintain the best heart health and effective blood circulation, it is crucial to meet the needs of the digestive system. To achieve this, it is advisable to consume certain natural oils and animal fats. The body's reactions to vegetable fats are interesting and negatively affect blood circulation. For instance, using sunflower oil triggers an unexpected and excessive production of amino acids in the pancreas. As a result, the body produces an excess of enzymes for the digestive system. This situation may start with symptoms of reflux and might lead to pancreatic cancer.

The food and drinks we consume in our diet notably affect the digestive system, which includes the liver, pancreas, and stomach. Keep in mind that the more natural and unprocessed your diet is, the greater the benefits it provides. Moreover, any issue in the pancreas induces feelings of unease and anxiety.

If you suddenly experience a strong desire for food, it may likely indicate a liver malfunction. The primary cause of this digestive problem typically resides within the circulatory or urinary systems.

When the kidneys do not develop properly, they fail to perform their filtration function effectively, consequently causing another issue within the digestive system. The interconnected hardware of our organs impacts our development throughout life, establishing the groundwork for our neural system. Further exploration of this topic will be undertaken throughout this chapter.

In addition to eating a healthy and natural diet, it is essential to engage in physical activities and sports. The type of activities significantly influence our overall health. Focusing on activities that promote overall body development, rather than intense or heavy sports, is important. People who build muscles through exercise but spend long hours sitting in front of a computer without using those muscles often experience problems.

The body covers muscle with protein, a process that can lead to sagging—a visible problem, and, at the same time, one that affects the quality of thoughts—a psychological problem. Our body expects the use of the areas we have strengthened. If these areas remain inactive, the body considers them unimportant. This condition eventually leads to episodes of depression and can rapidly disturb our well-being. As a result, if we engage in physical activities meant to develop our body, the outcomes must align with our lifestyle.

If we fail to include this development in our daily routine, we unknowingly subject our organs to a purposeless process. Regardless, our lives rely on these organs. For instance, contrary to common belief, having a well-defined six-pack may not symbolize health. It may even lead to digestive issues like constipation.

One may naturally ask, "How does constipation relate to software development?" Interestingly, those who experience frequent constipation are more prone to panic attacks. Contrary to the common belief, panic attacks are not only neurological disorders, but often stem from digestive problems and changes in internal digestive system pressure.

As our bodies undergo physical changes, we also reach new balances. Each balance we achieve propels us into the next stages of development. This development then leads us to a satisfying and healthy sexual life.

The outcomes reflected in our sexual life serve as an indication of the state of our physiological "hardware." Balance is closely linked to our sexual life, and if disruptions happen in this area, our organs might undergo problems. Sexual life is not solely about finding pleasure or satisfaction. Keeping a healthy and efficient sexual life improves effective decision-making abilities in one's social life. Individuals with irregular sexual routines often struggle with effective decision-making. This is a common problem that often doesn't get addressed because it is not widely acknowledged.

Every aspect of our body, including breathing, fatty liver, adrenal glands, pancreas, and even our feelings, affects each other. This signifies a hardware balance within the body where each part influences the next one. To maintain our body in this balanced state, we need to consider it as a whole rather than as separate parts. Therefore, when you encounter a health issue, it is important to examine it through the lens of the circulatory, digestive, and urinary systems rather than focusing only on a specific organ.

Besides all the factors mentioned earlier, a crucial sign of your development is the state of your skin.

Suppose you notice unexpected skin problems like a rash or itching, which cannot be linked to a fungal infection or classified as eczema or similar conditions. In that case, there is a strong possibility of a problem within your circulatory, digestive, or urinary system. Failing to address the underlying cause means any topical treatments administered to the skin will not fully rectify the problem.

Maintaining the balance of your body will enhance your self-confidence and fortify your hardware activities in the realm of software.

> The Spirit_

You may have read numerous books on improving spirituality or participated in spiritual workshops. In this section, we will explore the common elements of these spiritual studies. As highlighted in earlier sections, maintaining a healthy spirit is vital for achieving equilibrium between our physical bodies and nervous systems. We will delve into the reasons, benefits, and potential pitfalls of adhering to the standards of goodness as defined by your personal situations and norms. This concept of spiritual health transcends mere inner peace; it is intimately connected with our moral and ethical choices and behaviors.

As explained in earlier sections, a healthy spirit is crucial for maintaining a balance between our physical bodies and nervous systems. We will discuss the reasons, benefits, and possible mistakes of ethically and morally being good within the situations and norms you define. This spiritual health is not only a state of peace, it is also intrinsically linked to our moral and ethical behaviors and decisions.

An important point to remember is the reason behind one's desire to be good. If the decision to be good comes only from self-interest, you might not fully experience the processes explained in this book. This is because when we aim for goodness, we seek inner peace, desire, truth, and the motivation that arises from the journey.

We all should aspire to dedicate ourselves to life, meet our responsibilities, achieve success, keep our motivation, and find inner peace. If these motivations come only from our self-interest, we could compare that scenario to seeking survival without food or water. The human body depends on outside support, including sunlight, air, and many other factors for its survival. Even when we are not always conscious of it, these natural environmental conditions meet our basic survival requirements. We absorb these requirements and keep on existing.

Nature gives water, food, air, and other necessary things to us. Those giving others those gifts might not completely comprehend their impact on our being. The main idea is that nature is actively supporting us. Within this caring environment, we manage to live, endure challenges, and persist. Thus, if we perceive the peace, desire, truth, and motivation coming from nature, it might teach us a deep lesson. This lesson is known as "good for others." When one is capable of sharing nature's gifts with others, it significantly enhances their journey toward becoming a better person.

Setting aside your free time for others, using your savings to help someone without a home, inviting a friend to an enjoyable event, buying two tickets to a concert and giving one to someone who can't afford it, or engaging in activities with those who have limitations—each of those actions is meaningful. Those acts of kindness bring about positive outcomes, regardless of one's personal beliefs.

The point isn't about the exact action, but about actually doing something for someone else. Everyone aspires to be a good person in their areas of expertise. Whether it is a philosophical approach or a practical one, all efforts ultimately come together with a common goal. Certainly, the goal of the educational system is to equip individuals with the tools to lead a life characterized by peace, enthusiasm, groundedness, and motivation.

Those values come together under the term "goodness." Yet, when they are pursued solely for personal benefit, the quest often leads to failure. Hence, it is time for all educators to come together under a shared framework. No matter which ideology you take part in, the important thing is to share the knowledge you gain with others. Doing so will enhance your learning efficacy and speed up your transformation into a better person.

When you spread goodness to others, nature responds in kind, providing chances to find the inner peace you are looking for within yourself. Once you establish a deep sense of peace, it sets the mechanism of enthusiasm in motion. These mechanisms cannot be consistently and extensively acquired through any other method.

You cannot confront realities without enthusiasm, as the desire to go somewhere or do something serves as a means of comparison to the realities of life. Facing realities as soon as possible encourages greater motivation. However, starting this process with a self-focused approach might steer your journey in the wrong direction. When positive outcomes do not directly belong to you, they create a weight on you.

Lying to others always creates some level of tension in your life, regardless of your intentions. The functioning mechanism of lie detector tests confirms this idea. As a result, lasting relationships with others might weaken. Eventually, you could end up stuck in uncertainty or, even worse, feeling alone and left behind.

The body needs satisfaction, while the neural system desires peace. The balance between those, along with other processes, gives rise to the necessity of spiritual goodness.

As mentioned from the beginning, the central idea is not whether something is bad or not; the key is how much one can spread goodness and offer it to others. Instead of spending all your time on your computer, set aside time to help others.

Engaging with others will ignite your desires, enable you to view facts from a more realistic perspective because you observe different lives and their details, and ultimately lead to a life filled with more motivation. As you start on this path, you will discover that your inner peace becomes lasting, and your desires gradually come true.

> The Mind_

Maintaining the optimal health of your neural system is important for cultivating harmony between your body and spirit. While there is much to delve into regarding the nervous system, it is crucial to analyze topics quickly, particularly for developers living in this fast-paced era.

Despite the many topics connected to the neural system, they all eventually come together at a single point: perception. Grasping the essence of this touchstone is very important for analyzing later processes.

A primary outcome of the human neural system is the ability to perceive. Perception is the process of gathering occurrences, events, and their related phenomena and transforming them into coherent and meaningful situations. Rational thinking relies on perception. Without the ability to think rationally, the capacity to follow mental processes is limited, which could hinder career growth. That standstill can lead to problems with physical or spiritual health.

It is crucial to avoid conflating spiritual health with psychological well-being.

Chapter 7: /... / The Process of Perception and Perceptual Disorders

Events are the range of actions we experience throughout our life. Each action acts as a sign, showing that an event is happening or about to happen. A phenomenon is the expression of those actions. Unlike a word or statement, a phenomenon is an observable fact or situation that exists or occurs. However, perception goes beyond the literal meaning of a word. Every action matches with a corresponding word, and sometimes, you might be the one who expresses it. Perception is established when all these phenomena and events come together and work in harmony.

The process of perception begins at the point where an event and phenomenon come together. Considering the progress in modern technology, the traditional psychological definition of perception is no longer valid. Therefore, we should define the total of processes happening between events and phenomena as perception.

Phenomena originate from actions and sometimes these actions do not need words to be defined. No matter the circumstances, when events and phenomena come together, they should create a coherent and logical outcome. Thus, our perception—serving as the keystone of rational thought—will manifest with exceptional clarity and precision.

< / 1_ Confronting Anxiety >

When phenomena align with events, one's perception can be considered normal and within expected parameters. However, if an element of an event or a phenomenon either increases or decreases, changing the total value, it leads to a perceptual disorder. Therefore, any deviation, whether an increase or decrease in the total of events and phenomena, can lead to perceptual disorders in humans. A resulting condition that then appears is anxiety. Anxiety can become so strong that a person has trouble with everyday life, constantly feeling the need to double-check things. Thoughts like, "Did I write it right?" "Did I miss something?" and other similar worries can make the anxiety worse.

> Exercise_

A practical solution to address anxiety involves participating in exciting activities, like riding a rollercoaster. People who experience strong anxiety often avoid participating in activities like riding a rollercoaster because they evoke stories that might lead to death. Therefore, involvement in sports and activities that boost adrenaline and bring excitement is important for individuals struggling with anxiety.

By actively participating in adrenaline-boosting sports and social activities, individuals realize that certain events and phenomena are perfectly normal and actions performed by many are ordinary. In this process, they learn to simplify events and phenomena. Removing factors that trigger anxiety allow the brain to comprehend events and phenomena better and to fictionalize them without exaggeration.

< / 2_ Overcoming Impassiveness >

On the other hand, if the aggregate of the event and phenomenon falls below the expected level, the resulting state is called "impassiveness." In this situation, a person remains uninterested in the outcomes of events, even when they are directly affected. For developers, impassiveness is as significant a threat as anxiety, and it can lead to job loss and limit personal growth.

> Exercise_

As mentioned before, a practical way to address impassiveness includes gaining a better understanding of poverty. You could try spending a night outside, like the homeless. Comparing your impassive life to the possible lack of shelter might lead to a renewed sense of purpose, making you face the realities of everyday life more directly.

Perceptual disorders, even though they are often overlooked, are frequently more common than stress in the field of software development. Permanent stress, different from its short-term version, arises from perceptual disorders.

Based on information from psychiatric experts, disorders linked to permanent stress happen less often, leading to the lack of adequate focus on perceptual disorders. However, it is important to recognize that perceptual disorders are the foundation of all mechanisms of permanent stress.

Individuals' stress levels change based on their attitudes toward events and related phenomena. Regulating perception can help reduce stress, leading to a more balanced nervous system. Engaging in various hobbies may also help reduce stress levels. While these activities can provide relief, the recurrence of stress from the same triggers may lead to permanent stress. It is important to remember that every persistent stress pattern is rooted in a perceptual disorder.

To regulate perception, it is necessary to understand events and phenomena clearly. Those who have reached this level of understanding are often described as mature individuals. Individuals who mature early often develop a life without perception-related problems from a young age. This arises from their exposure to a diverse range of events and the repeated experience of the phenomena associated with them. This process facilitates the realization of their true perceptual capacity, leading to reduced stress levels.

Maturity isn't only about getting older. The significance individuals attribute to events and phenomena, coupled with their comfort in confronting the reality of death—a profound concern for some—plays a role in regulating their perceptual level and algorithm. At this point, selective perception loses its significance, enabling optimized responses to events and phenomena. As a result, perception becomes regulated, leading the individual to attain a state of maturity.

Having explored psychological aspects, let's turn our attention to sociological aspects.

Software development profoundly impacts our social lives, but many people do not realize it is a sociological process. As a developer shapes the social lives of others, their actions also define their position in society. Interestingly, developers often acquire the reputation of being introverted individuals.

In the past, communities considered people who were overly outgoing and friendly as crazy or strange. Today, people see developers as individuals who spend their days focused on their computers, dedicating their lives to writing code. People may hold various viewpoints about developers, yet developers must recognize their role in the sociology of programming to understand their existence better.

Before the technological revolution that strongly impacted our lives, closer and more intimate methods of communication defined human communication, requiring being physically close and having direct conversations. People coming together, sharing thoughts, and exchanging ideas led to the emergence of social gathering places like cafes and restaurants. Those public spaces and the online world have exploded in the last fifty years. Progress in software technologies enabled the creation of modern communication methods suited to the current time's needs.

The older, more personal ways of communication were time-consuming, leading to a shift to social media platforms. That caused society to adopt a new way of communication based on time management. While this change has many positive aspects, it is important not to forget how it has impacted our relationships based on writing and personal communication. Many social scientists see this shift as one of the key social problems of the last century.

At this point, it is important to understand that humans are undergoing a shift from a concept of empathy to one of inquiry. Close and intimate communication requires you to develop the empathy to comprehend the feelings, thoughts, or desires of others. If you cannot comprehend someone or do not share similar values, you should try to grasp those values. However, this approach has become less important nowadays, as the ideas you come across are probably not entirely new to you.

Even when faced with entirely new concepts, the social foundation provided by the internet helps in discovering their meaning. Developers mainly exist within this realm, where they shift from an empathetic perspective of a student. Developers, inherently more focused on learning than empathizing, leap to the next stages. That could explain why they might seem disconnected from Gen Xers. But this does not mean developers are naturally not social.

Chapter 7: /... /... /... / The Sociological Impact of Software Wars and Social Media

Before the widespread impact of technological progress in our daily lives, mainly the circumstances of economic and political difficulties influenced people. The time spanning before and after the Cold War greatly formed individuals' views on the war in various economic and social aspects.

The growing field of software led to "software wars." Politicians started using platforms like Twitter to make statements and wield influence in the realm of politics—a previously unimaginable notion. The start of Donald Trump's use of social media platforms brought about new diplomatic concepts, leading to significant sociological changes in society.

The sociological structures that supported financial and political conflicts required people to develop resilient characteristics. There was a widespread, though sometimes hidden, feeling of obligation to resist among individuals, signifying their pushback against systems they found unjust or oppressive. In the past, this resistance was often a subtle act of resilience due to the risks of open opposition. However, in the present era, the advent of social media platforms has allowed people to openly share their preferences and views, clearly presenting various arguments and ideas. This shift indicates a transformation from covert resistance to more overt and public forms of expression and advocacy.

Today, there is no need to resist anything. When we are certain of the accuracy of our choices, we can proclaim them to the world without encountering resistance. Physically, nothing can stop us; only the "likes" or "dislikes" of those who understand our statements can influence us.

In the era before the digital age, core values held immense significance within societies. Consequently, imparting moral and religious education was essential for communities to uphold those values.

Changes in everyday social structures have led to the creation of new, evolving beliefs. In today's society, altering religious morals drives an increasing number of individuals, including developers, to seek their true selves. The question naturally arises: You want to be yourself, but who do you want to be yourself for, and in relation to what? It is here that a social issue comes up.

As time goes on, people will long for the times when they could gather around a table and take part in honest conversations, reflecting the customs of the past. In this social environment, people will try to reconnect with others on social media who show traits that are similar to or different from their own. Certainly, this transformation in society will lead to a new development: interdisciplinary software development experts.

The mediators of our future will not be diplomats or philosophers, but developers from various fields. It is possible that, at some point, lawmakers may seek developers' opinions on their decisions based on collected data. Developers will act as mediators in this situation, shaping fundamental values during this interdisciplinary transformation. Recognizing our imminent proximity to this emerging reality, we might need to set aside our self-awareness.

It is important to remember that we cannot change the fundamental values inherent in nature. Neglecting this truth and trying to change it might evoke extraordinary, perhaps even supernatural, reactions. When confronted with the supernatural, humanity will always be crushed.

If our goal is to live a peaceful life, we need to be cautious when using the benefits of technological advancement. Regardless of your basic values, you can successfully navigate this sociological change in the most positive way without compromising those values. This positivity will give you the responsibility to advise politicians and government groups. Therefore, it is important to never ignore the values of your present society without confronting the possibility that you might someday be involved in changing them.

Chapter 7: /... /... /... /... / Psychology and Characteristics of Software Developers

Undoubtedly, developers living in this social environment have a distinct psychology compared to others. This is due to the rise of a new psychological perspective brought about by this interactive setting. Considering this situation, it will make sense to support the idea of having more psychologists who specialize in working with developers. This suggests a gradual increase in the count of occupational developers, leading to a future where psychologists might find themselves using applications developed by medical software designers.

Groups of developers that concentrate on making products for their professional peers rather than trying to understand every occupational group will inevitably lead to new fields of study.

In classical psychology, an individual's quality of life is measured by their energy and interaction. If their energy and interactions are unusually high, a person might be seen as manic, while those who act more introverted and seem embittered might receive a diagnosis of depression.

To define "normal," a midpoint between the states of mania and depression was chosen. However, it is currently not possible to classify developers as manic or depressive when evaluating their psychological landscape. Because they exist outside the norms of typical psychological life, we have a responsibility to address their unique psychological needs distinctly.

Now, it is possible to classify individuals into three types: positive, plus-positive, and minus-positive. To tackle these disorders, it becomes crucial to comprehend the problematic aspects faced by each type and then create a customized treatment method accordingly.

</1_ The Positive Software Developers >

Developers belong to the group of positive individuals. Those who can impact the lives of others, whether positively or negatively, are considered positive individuals. Modern psychological research suggests that the positive type is closely linked to normal conditions, but it is not defined in relation to manic or depressive states. The senses of positive individuals function properly, distinguishing them from both minus-positive and plus-positive individuals.

</2_ The Minus-Positive Software Developers >

Individuals within the minus-positive group have a noticeably active role in daily life, often seeing themselves as the VIPs and stars of the bunch. Such effective individuals can guide their group, finding joy in performing beyond their assigned duties. These individuals are effective at criticizing someone else's work while simultaneously trying to complete it. While initially presenting as overwhelmingly positive, this perspective could eventually result in feelings of inadequacy due to others' elevated expectations. This could lead to the development of personality disorders.

Active individuals succeed not only in being highly effective in their professional and social environments, but also in shaping their relationships. They decide the role others play in their lives and compel them to adjust to their surroundings. As a result, they may exhibit both oppressive and despotic inclinations.

Individuals who consider themselves as highly effective and active in daily life often see themselves as stars. The idea of being a star, or the efforts to become one, defines the minus-positive individuals. As they achieve more and make progress, these individuals become more and more certain of their belief of superiority over others. As shown, the word "belief" is crucial in diagnosing all three types because minus-positive individuals rely on their own belief systems, which can deviate from reality.

</3_ The Plus-Positive Software Developers>

Plus-positive individuals represent a different aspect. When comparing them with minus-positive individuals, a different conclusion becomes apparent. Unlike the acknowledgments from others, plus-positive individuals see themselves as unsuccessful. Some might perceive the negative-positive traits I just mentioned as exhibiting manic characteristics, while the positive traits may showcase characteristics associated with depression. However, these attributes reveal inner states rather than outward behaviors usually associated with manic or depressive individuals.

Putting software developers into categories is a tricky job, as they either withdraw from or keep away from addressing the needs of their inner worlds. Diagnosing plus-positive individuals requires focusing on specific keywords.

The first keyword is "failure." Individuals in the plus-positive group inherently embrace failure. While a person in a depressive state perceives failure as a problem, an individual from the plus-positive group readily embraces it. No matter how much you attempt to show such a person their achievements, they will remain unconvinced.

The second keyword is "despair." The feeling of despair experienced by plus-positive individuals often leads to ignoring potential solutions. These individuals believe those around them fictionalize their surroundings, leaving them with a sense of responsibility solely for their existence within this fictional world. They regard despair as both a necessary and sufficient condition. Therefore, they persistently reject solutions.

The third keyword is "solitary." Software developers who align with the plus-positive category perceive themselves as solitary. Even when they are not lonely, they actively seek solitude.

For individuals diagnosed with depression, loneliness is a state in which the people they desire to be with are not around. However, plus-positive individuals perceive themselves as solitary entities, as if they have no parents, siblings, or anyone who loves them in their lives. It is as though their existence began in the middle of emptiness and the absence of human presence. This is a why many developers today struggle with problems in forming connections with their families. This mindset is more about accepting desolation than desiring solitude and involves giving up the chance of having companionship.

> *The difference between the Positives_*

Notice that belief is more dominant in minus-positive individuals, whereas rejection is more clear in plus-positive ones. Developers who lean toward either belief or rejection might consequently experience a reduction in certain senses. Individuals diagnosed as manic or depressive assess their losses based on personal situations and subsequent outcomes. However, such an approach is not suitable for software developers. Therefore, deficiencies in specific senses display characteristics that lead to the plus and minus in positive types.

The main reason minus-positive individuals have firm stances in their belief systems and the inability to trigger motivational factors to influence their bodies is mainly due to their loss of a sense of taste and smell. As a result, minus-positive individuals teach themselves how foods should taste or smell. For instance, when a minus-positive person eats a mango, they have already predetermined its taste and scent in their mind. No matter the actual quality of the mango, they would perceive it as enjoyable. That disconnection from their sense of smell and taste highlights their minus-positive state, showing a divergence from their sense of reality.

In general, the feelings of failure, despair, and desolation felt by plus-positive individuals who reject certain aspects of life are embodied as a loss of sensation, especially in the sense of touch. This loss stems from their internal rejection, translating into a tangible, physical disconnect. They perceive temperature based on their perceptions. It is not unusual to see plus-positive individuals wearing a t-shirt in freezing weather or a sweater in scorching weather. The clothing choices of individuals, especially those in Silicon Valley, reveal their respective minus- or plus-positive traits. As a result, we observe that plus-positive individuals, inclined to reject, also disconnect from reality.

Whether they lean toward the minus-positive or plus-positive side, these individuals are prone to becoming disconnected from reality. If both types move closer to reality, they will also move toward a more positive state simultaneously. At this point, grasping the true essence of reality, a concept that exists independently and is not defined by comparing it to something else, becomes extremely important.

A common problem experienced by modern software developers is their tendency to compare themselves with others. This is a primary factor that contributes to the categorization of plus- and minus-positive types. When developers compare themselves to exceptionally successful individuals in Silicon Valley, they feel disconnected from reality. That separation from reality, linked with their social lives, family interactions, and even their typing style or sitting position in front of their computer, speeds up their movement toward a plus or minus-positive state. To stop comparing one reality with another, it is essential to acknowledge that reality is, in fact, real.

This concept can be best understood by explaining it like this: if someone is unfamiliar with the taste of water, they will not be inclined to develop software about water. It requires understanding its feel, smell, color, and other qualities.

Regardless of your specialization in development, it is important to understand the realistic aspects of the subject. For example, if you are involved in making entertainment software, it would help to have some experience or involvement in the entertainment industry. If you do not recognize the genuine aspects of entertainment factors in your work, you will keep comparing your ideas with others', resulting in creations that lack realism.

Remember that you live in the real world, not a digital one. Your feet remain grounded on this earth. You eat, drink, smell things, and feel touch. You need love and affection. These things are inherent to your nature. Separating yourself from these natural qualities and facing more problems linked to the positive and negative states without dealing with them will eventually require the help of a psychologist who specializes in addressing the psychological concerns of developers. As mentioned before, this particular specialty is currently not present; however, as research and studies advance, there will surely be more psychologists who focus on this and will create a new profession.

> Exercise_

To reconnect quickly with reality, engage in activities that directly interact with natural elements like soil and water. Try making pottery or sculptures with your own mix of soil and water. Interacting with these elements will enhance your senses and help you understand life's realities better. Furthermore, while fire and air also hold significance, when considering the psychological perspective and the foundational role that soil and water play in human composition, they can be considered the most realistic, natural, and readily accessible elements.

Engaging with soil and water gradually helps your brain connect with reality, guiding you toward a positive state of mind.

Chapter 8: / Interdisciplinary Connections

The interaction between software development and other scientific fields undeniably promotes significant advancements while impacting a developer's daily life. Considering the powerful impact of the connections among different scientific areas on human life, reviewing and examining these fields with more accuracy and depth is crucial.

Think about the relationship between geography and software development. The significant strides made by humanity in its pursuit of software development, leading to positivity, are something we cannot overlook.

Is positivity a process solely pertaining to humans? Definitely not. You can notice positive currents in the world's orbit when examining geographical conditions. We also know that Earth's electromagnetic, gravitational force creates negative currents due to certain natural circumstances. Learning about these occurrences can offer you, as a developer, a new perspective, and it could help you find useful areas in your everyday life.

Remember that this book serves as a trigger, pushing forward your explorations and advancement in software development. Without a doubt, being close to areas with a positive current is important to make the most of the Earth's electromagnetic force and its positive effects. Positive effects ensure quicker recognition and solving of problems. People know that living in areas with positive energy flows ensures a more comfortable and easier life compared to those without such benefits. In cities along positive current lines, such as New York, Beijing, and Istanbul, people show greater productivity while feeling less tired.

Negative current is slow and decreases the speed of others. When examining Earth's rotation, we observe variations in rotational speed, with regions near the poles turning more slowly and those around the equator spinning faster. That shows the influence of geography on people.

As our comprehension of Earth's geography and intrinsic energies has developed, another area where our understanding continues to expand is the interplay between hardware and software within the realm of technology. Looking at the situation from a different perspective, it becomes clear that present developers have not fully used the potential of hardware.

There is an opportunity for us to use the hardware we already have more effectively with the help of the software we produce. We are currently entering a phase where it is very important to improve the efficiency of software to get faster results. In this process, while hardware maintains certain standards, software has the potential to create efficient and successful outcomes, especially in the context of dominant chipsets.

In Chapter 7, we compared our physical body to hardware. As humans, our understanding of the intricate processes and mechanisms happening within our bodies is still restricted. Just like the unexpected benefits of physical activity become apparent, the positive impacts of software on chipsets will also confirm that hardware does not have to progress at a similar speed. We are entering a time when software and hardware developers must work together in harmony with positive flow. If someone lives or spends time in areas with these positive currents, they will certainly understand the message we want to convey.

The global positive currents reach their highest point on sharp mountains by the sea, as well as on tall, pointed mountains in deserts. Volcanic mountains, in particular, have notably higher positive current values. You might be far from these and similar positive areas. In that case, you can enhance positive effects in your home using scents and natural stones. For instance, houses with natural salt rocks are less affected by negative currents.

Avoid choosing marshes, small valleys near streams or creeks, small or dry lake areas, rocky landscapes, and treeless plains for your vacation, as these areas have the highest levels of negative effects. Selecting geographically high-efficiency places for your vacation and absorbing the positive information present in these areas can refresh your outlook in the field of software, thus helping you make the best use of chipsets.

Chapter 9: / Rethinking the Past to Innovate the Future

All technological products we have today come from economic, political, diplomatic, and social factors, as technology essentially originates from human needs.

Our needs originate from points where certain things become necessary. Times of war, conflict, new advancements, economic developments, and political transformations often coincide with recognition for individuals that they need something different. We should not limit the historical analysis of this sociological structure only to the scope of anthropology. As we move through the stages of software development, it is important for us to cultivate a better grasp of historical changes and continuously reassess them.

Creating a library designed for developers of historical books would be beneficial. In this library, we could develop a new algorithm that emerges from sudden changes in medical science, sociology, literature, and similar areas. For instance, consider Roman cisterns not solely as solutions for water needs, but also as strategic constructions for emergencies. Despite the great power of the Roman Empire, the architects built those cisterns with the foresight of providing cities with water during potential war scenarios. At the same time, they used this water source in baths and public bathhouses. Therefore, the importance of cleanliness in daily life, seen in the famous Roman baths, shows how the religious beliefs of that time might have promoted the concept of cleanliness.

The Romans introduced the practice of baths and public bathhouses in response to labeling barbarians as dirty. Historical records from Ancient Greece also indicate that government affairs and customary laws frequently excluded those seen as unclean.

Is today's society any different? No, it is not. We often label individuals attempting to corrupt things as "dirty," and subsequently, we strive to introduce a "clean" solution to remedy their actions.

In the current political atmosphere, there is a strong push toward specific requirements. Governments expect developers will shift their focus to the defense sector. However, the world needs software that meets personal needs and fosters social unity instead of just focusing on the defense industry.

Ancient philosophers followed a path similar to that of software developers in the past. They supported their scientists and shared their gathered knowledge and data. Their influence came from their expertise in history, mythology, geography, and all the mentioned values and qualities this book highlights. Modern developers lack the backing of philosophers or individuals knowledgeable in this arbitrage culture who view history through such a perspective.

It is crucial to re-articulate the fundamental reasoning behind this philosophy. Without fresh opportunities for software developers, there is a common belief that they will focus only on improving their abilities for their benefit rather than using them to shape the world. However, throughout history, individual talents have often been left unfinished because of the limitations of human life.

As we can observe today, modern, well-known software companies frequently sacrifice certain aspects or elements in their pursuit of progress such as user privacy, ethical considerations or quality assurances. For a strong understanding of this intent, it is crucial to comprehend the underlying system. A mere brief investigation does not offer enough grasp of this understanding. The interplay of these structures throughout history serves as a foundational basis for creating apt metaphors for new historical processes that will emerge from future software developments.

> Software's Societal Impact and Legal Changes_

The legal dimension of the software world mainly centers on safeguarding the financial rights of software applications. However, as we step into a new era of AI-driven development, a pressing need emerges for a reimagined legal perspective on software. In this changing legal perspective, our influence on the lives of others will gradually expand.

Experts have examined the effects of software technologies on children in a more comprehensive manner. As the popularity of games increased, many aspects of children's lives changed. Researchers carefully examine these shifts, and their discoveries will pave the way for new legal areas.

Furthermore, the legal system will experience significant changes. It will need to establish a separate area focusing on privacy concerns due to the growing impact of software developers on people's lives. At this stage, the software industry might have to create its own legal structure, not to change people's lives but to understand the legal consequences of software's influence on individuals.

When assessing different groups of professionals, we measure what we refer to as "law" by the equivalents of that profession in everyday life. The lack of understanding of the consequences of actions taken by software developers persists. Currently, the software world has difficulty expressing and explaining its essence effectively. However, this ongoing change might reveal fresh opportunities for us.

The realm of the software world is increasingly interconnecting with the domain of law. As AI continues to advance, there is the potential for it to replace lawyers and prosecutors.

It is worth noting that research on AI, though not always highlighted, covers a wide range of legal matters, from individual lawsuits to corporate disputes. Furthermore, it is important to remember that the fields of advocacy and prosecution are relatively new compared to the well-established judiciary. Many people are still unaware of how AI might enhance this domain.

However, it is clear that as the software world advances in legal matters, the roles of prosecutors and lawyers will likely disappear. Considering the interaction between developers and the legal system, experts hold the belief that developers could play a vital role in reshaping the law, with this transformation potentially unfolding in the near future. However, in this rapidly evolving era, we can expect that the time required for such progress will be relatively short.

> Software and Economic Evolution: Cryptocurrency_

In analyzing the interplay between the software world and legal matters, it becomes apparent that financial elements also carry considerable significance. With continued growth in the software world supported by financial investors, we might need to reconsider our views on cryptocurrency. The lack of production in cryptocurrency is likely to transform into a system based on production, guided by the values established by the software world. This change will lead to significant transformations in the cryptocurrency field and legal processes.

Software developers will play a key role in bringing together interdisciplinary fields with governments producing cryptocurrencies. Their contribution will extend to aiding governments in comprehending the effects across various disciplines and throughout the auditing stages of cryptocurrency production. At this point, production-based cryptocurrencies will emerge. Users will give feedback on software applications, and developers will earn a share based on the popularity of their products. We will restructure the system to ensure that developers contribute to a more beneficial process.

When examining the connection between agriculture, medicine, and production, the financial advantages derived from this process will establish a strong and genuine economy. Maintaining balance and discipline between the output of a potato farmer and the software developed by a company is crucial. As a result, production-based cryptocurrency will rise as a shared value within the production field, signifying a significant shift for software developers.

Software developers' efforts to produce cryptocurrencies will collapse from top to bottom within the next ten years. Instead, the focus will shift to production-based cryptocurrencies. As a result, it will be best for developers and software companies to focus on production and embrace production-based cryptocurrencies.

What is of utmost importance is to refrain from creating currency for assets that have not yet been produced. With the collapse of the bubble system in the capitalist interest economy, users will replace investors. Their support and contributions will surpass those of traditional investors. Investors will value the interdisciplinary elements created by cryptocurrencies. This measure will result in stability, and central banks will produce digital currencies. As a result, the influence of thc software world will become undeniable.

In the past, individuals who extracted gold and silver played a crucial role in producing money, placing them at the top of the worldwide economy. Likewise, this principle is evident in the current cryptocurrency landscape. To prevent illegal actions and ensure government oversight, developers should control these processes and be supported by users. All these factors indicate that developers will not only play a significant role in economics and law, but will also occupy a central function. In the future, many areas, especially the bureaucracy, will work within a system directed by software developers.

> The Software Developer's Role in Legal and Economic Spheres_

Keep in mind that unproduced assets do not equate to money. Consequently, non-existent events will not correspond to legal realities either. In the near future, AI will replace unethical lawyers and prosecutors who endeavor to validate faulty evidence. Similarly, we can say that cryptocurrency will change along with the evolving AI.

Throughout history, legal and economic progress has developed in parallel. Every economic initiative risks failure when economic advancements surpass the existing legal system. Support for legal changes highlights the fundamental requirements for a shift in the software field, thus enabling substantial economic advancement. However, it is also incontrovertible that individuals studying these subjects often fail to recognize the interrelated outcomes across various disciplines.

A software developer should possess the clarity and quality to write software in the fields of law and economics, bring different disciplines together with the software they create, and have foresight on these matters. At this point, what is important is to adapt to the changing world and maintain a foresight approach when dealing with situations within this context of clarity.

> Beyond Code: The Future Fusion of Software, Law, and Biology_

We have discussed the primary matters in a developer's life, but it is not within the scope of this book to cover all aspects of their future. We covered the beginning of their journey, their current position in our daily lives, the extent of their impact, and the essential changes a developer must make to create meaning.

In the next book, we will explore the emerging waves of software development that will impact not only the field of law but also the field of biology. After all, a careful examination reveals an unexplored area: the realm of genetics.

Research conducted by genetic engineers and scientists is in its early stages. Activities such as cloning a sheep, growing an embryo in a surrogate mother's womb, and other examples were mainly related to medical sciences. However, when thinking about future AI equipped with the ability to make choices and think, it is clear that a group of software developers will focus on living things. Therefore, since the beginning of the book, we have mentioned nature, history, geography, and the necessary steps for better understanding ourselves. Each step signifies a long adventure we will begin. When these statements come together to form a cohesive whole, this book will only serve as the first step of that journey. By taking its initial steps, you will be among the pioneers who successfully embark on this journey.

During our journey through the remarkable realm of software development and its profound implications, we have traversed diverse landscapes of knowledge, inspiration, and innovation. This odyssey has not only illuminated the pivotal role of software developers in shaping our rapidly evolving world, but has also provided glimpses into the boundless future that awaits those who eagerly dare to explore the uncharted territories of technology.

Throughout these pages, we have unearthed the roots of software development, tracing its lineage from the ancient philosophers who sowed the seeds of computational thinking to the modern-day pioneers who have harnessed the power of algorithms to change the course of history. We have witnessed the symbiotic relationship between technology and humanity, where each advancement begets new possibilities and challenges that demand our collective wisdom and ingenuity. As we stand on the precipice of tomorrow, it becomes evident that the horizons of software development extend far beyond the confines of code and programming languages. The software developer of the future is poised to become a multidisciplinary maestro, orchestrating symphonies of innovation that harmonize the realms of law, economics, biology, and beyond.

Our exploration has taken us from the birth of AI to the potential fusion of technology and genetics, painting a vivid picture of a world where the boundaries between the digital and the biological blur. It is a world where software developers will be the architects of not just software systems but the very fabric of life itself. In this grand narrative of the software developer's evolution, we find a call to action—an invitation to embrace change, cultivate a holistic understanding of the world, and harness the infinite potential of our craft. The challenges ahead may be formidable, but they are also the crucibles in which we forge the tools to shape a brighter future.

So, as you embark on your journey as a software developer, remember that you are not merely a coder, but a visionary, a problem solver, and a catalyst for transformation. As you code the algorithms of tomorrow, do so with the awareness that your work has the power to transcend boundaries, bridge divides, and bring about positive change in the world.

The future beckons and you are at the forefront of this exciting adventure. Embrace it with open arms, for in the boundless realm of software development; the possibilities are as limitless as the horizons of your imagination.

Go forth; your journey has only just begun.